The
Information
Imperative

The
Information
Imperative

*Managing the Impact of
Information Technology on
Businesses and People*

Cyrus F. Gibson

AND

Barbara Bund Jackson

With contributions by
Thomas H. Davenport, Gary K. Gulden, Tauno J.
Metsisto, David G. Robinson, and Thomas J. Waite

Foreword by
F. Warren McFarlan

Lexington Books

D.C. Heath and Company/Lexington, Massachusetts/Toronto

Library of Congress Cataloging-in-Publication Data

Gibson, Cyrus F.
The information imperative.

Includes index.
1. Business—Data processing.   2. Business—Communica-
tion systems.   3. Information technology.   I. Jackson,
Barbara B.   II. Title.
HF5548.2.G496   1987   658'.05   85-45797
ISBN 0-669-12338-2 (alk. paper)

Published simultaneously in Canada
Printed in the United States of America
Casebound International Standard Book Number: 0-669-12338-2
Library of Congress Catalog Card Number: 85-45797

The paper used in this publication meets the minimum requirements of
American National Standard for Information Sciences—Permanence
of Paper for Printed Library Materials, ANSI Z39.48-1984.

88   89   90   8   7

Contents

Table and Figures

Table

Figures

Foreword

No technology in the past two decades has as profoundly impacted the operations of as many organizations in as many industries as information technology. The combined impact of data processing, telecommunications, and remote office support devices have left few organizations untouched, with the future looking even more turbulent. Here are just a few of the major changes:

- Corporate organization structures have been profoundly altered, with decision-making activities being both transformed and moved to either a more centralized or more distributed approach in different settings. Powerful new management control systems have come into existence, allowing corporations to more effectively align their incentives with their operations.

- New linkages and services to customers and suppliers have emerged—in some cases, to the betterment of both parties, and in others to the betterment of one.

- The boundaries between industries have been redrawn as technologies blur decades-old structures. At the same time, new firms have emerged to service various segments of information systems (I/S) support.

- The pace of organizational life and decision making has dramatically quickened. Processes that used to take several weeks have been collapsed to a day and in some cases to a matter of minutes or less. Nonadapters to the new standards of this changing world have been driven from the marketplace.

- Finally, relationships between individuals, departments, and companies have been permanently altered. New divisions of tasks and new types of work have emerged, often with great pain.

Leavitt and Whistler, in their great *Harvard Business Review* classic of 1956, "Management in the 1980s," foreshadowed many of these things. As late as 1980, "Management in the 1980s" was cited as an example of flawed vision which never came to happen. Only in the past seven years did its full reality emerge in concrete form. Retrospectively it is clear that while their vision was unclouded, they dramatically underestimated the types of technology needed to execute this vision and the enormous problems of organizational transformation and change in management which had to take place for it to be achieved.

More important for the reader, the next decade will see the pace of I/S technology change continue to accelerate and knowledge on how to construct and manage applications deepen. A decade from now, the applications and technology of the late 1980s will look as primitive as the applications and management practices of the late 1970s now look. This book, *The Information Imperative*, transmits this message to managers in a powerful and readable way and gives them a framework for thinking about how these issues impact their lives and what they can do to respond positively.

The Information Imperative is not a theoretical book, but is instead a powerful, hands-on treatment from people who have been there. Yet, underneath this practical treatment lie strong conceptual underpinnings. The book's pragmatic generalizations rest on a firm foundation of the latest field research. *The Infor-*

mation Imperative provides a holistic view. Rather than starting with a particular technology or application and exposing it to detailed microanalysis, it begins by examining I/S's impact on the organization as a whole. Successively it then talks about I/S's impact on organizational units, and then its impact on the life of the manager. No attempt is made in this book to define a world where every manager uses a terminal. As a separate topic, the implications of technology on the I/S department are defined and practical guidance is given as to how the I/S department must be transformed in the coming years to remain relevant.

The central theme of *The Information Imperative* is the necessary and creative partnership that must exist between managers, who understand the corporation's products, structures, and culture, and I/S technologists, who understand what today's and tomorrow's technologies can do. Only through the creative management of that partnership can the organization's long-term future be assured. Both managers and technologists can profit from a shared reading and discussion of the issues in this book. Failure to establish and manage this partnership would mean long-term and unhappy consequences for the organization.

F. Warren McFarlan
Professor of Business Administration
Harvard Business School

Preface

This book would be incomplete if we did not acknowledge an important source of our thinking and insights: our clients. Since 1969, we have been working with major corporations to put information and information technology to work to achieve significant, lasting business results. Our greatest value has come from taking ideas at the cutting edge of experience and translating them into practice. All along the way, we have remembered one important principle: you learn the most by listening.

And so we've listened—a lot—to our clients. Without them, these ideas and approaches would exist only in a vacuum. With them, the ideas have blossomed and matured and made a real difference. We're grateful for what we've learned in partnership with our clients and for the trust they have placed in us.

Thomas P. Gerrity
Chairman and Chief Executive
Index Group, Inc.

Acknowledgments

The authors wish to thank the following for contributing their thinking and experience to this book: Michael Hammer, Patricia T. Kosinar, Walter J. Popper, James E. Short, Steven A. Stanton, and John M. Thompson. Thanks also to Leonard Cohn of Monsanto and Richard Koeller of TRW, who reviewed early drafts. And special thanks to Diane Wilson and Katie Crane, who assisted with the writing and rewriting of many chapters. Finally, we'd also like to thank Thomas J. Waite, Director of Marketing at Index Group, whose contributions to content and style, as well as even-tempered persistence, helped move this book from an idea to reality.

Introduction

For the past forty years an information technology revolution has been taking place that continues today unabated. It is nothing less than a revolution to have gone from ENIAC, the 1946 computer that weighed thirty tons, had eighteen thousand vacuum tubes, seven thousand resistors, six thousand manual program switches and took two days to set up for one program, to today's personal computer, which weighs less than ten pounds and can fit into a briefcase or on a desk, is made entirely of microscopic semiconductors, and is capable of running a variety of programs in seconds. The memory capacity of a single chip and the transmission speed of telecommunications media (to name only two key measures) are doubling and tripling every three to five years, while unit costs are plummeting. Today's hardware, software, and communications systems are dramatically different in both price and performance from their 1946 counterparts. We expect this trend to continue and to accelerate. We have no doubt that *revolution* is an appropriate label for the rapid changes and growth in the information technology industry.

What impact does the revolution have on businesses and their managers? How will it affect business tasks and individual careers? Information technology enthusiasts often predict and promise more than the technology can deliver, while skeptics argue that the proponents' rhetoric is dangerous. We believe that the skeptics are correct insofar as they point out the difficulties in realizing the benefits of the information technology revolution. We also believe, however, that the enthusiasts are correct in understanding how great the benefits can be if the difficulties are overcome. Most important, we believe that those difficulties *will* eventually be overcome.

> Sooner or later, every industry, every firm, and every career will be surrounded by, dependent upon, and required to respond to information technology and its implications. The only question is how and when.

The information technology revolution creates great opportunities. To emphasize that fact, we could have named this book *"The Information Opportunity."* We did not, however, because that title makes response to the revolution sound voluntary or optional. We believe instead that response is obligatory and, moreover, that the need to respond is urgent. The potential of the information technology revolution creates *the Information Imperative:* The urgent, obligatory need for businesses (and individuals) to respond to the opportunities and competitive threats created by the revolution in information technology and the resultant huge increase in the availability of information.

How can managers respond to the Information Imperative, and when should they do so? Most managers simply have no experience, training, models, or working concepts to help them. To paraphrase the Cheshire cat's advice to Alice: when you don't know where you are going in the face of the information technology revolution, any road can get you there. Managers need help in charting a course in the face of the Information Imperative. This book provides that help, focusing on the "how" and "when"—how executives and managers can exploit information and information technology to achieve their business and personal goals and when to seize opportunities to gain competitive and career advantages. The book does not offer a detailed recipe for response. Instead, it presents a basic approach for meeting the Information Imperative, first for understanding it and then for selecting and implementing an appropriate response.

In chapter 1, we discuss the Information Imperative. Our goal is to increase the reader's ability to perceive the Information Imperative in his or her own career and business and to provide an intellectual foundation for subsequent decision making and action.

Chapter 2 introduces the Benefit/Beneficiary Matrix, a framework for sorting and categorizing opportunities for using information and information technology. In our consulting practice at the Index Group, the Benefit/Beneficiary Matrix is a tool for assessment, diagnosis, and planning. It underlies almost all of our work in information technology, in diverse industries and businesses. For many of our clients, it provides the senior manager's pocket checklist of questions to ask and issues to address regarding information and information technology. It highlights the fact that successful response to the Information Imperative often requires substantially more than incremental improvement to a business's activities; instead, success often requires fundamental redefinition, or *transformation,* of businesses, functions, and even personal careers.

Throughout chapters 3 through 6 we describe the impacts of the Information Imperative on different beneficiaries: the business as a whole, the functional unit, executives, and other individuals.

Chapter 3 examines the Information Imperative from the perspective of the business as a whole, discussing the threats it poses and possible responses. We believe that all managers, particularly senior line managers, need to understand and act on the threats and opportunities. In chapter 3 we review the major forces challenging most of today's organizations: changing customer relationships, channels of distribution, competition, organizations, and technology. We argue that successful responses involve three steps: understanding, designing (actions and processes), and implementing. Finally, we suggest three information-based solution categories that have the potential to increase efficiency, to improve effectiveness, and to transform the business as a whole: (1) redefining the marketplace offering, (2) restructuring internal and external connections, and (3) redesigning business and/or management processes.

Chapter 4 considers the Information Imperative for the specific business function or department. We argue that the challenges for the function are primarily the same as those for the business but that the function may feel special pressures for cost control and for coordination. Successful responses involve understanding, designing, and implementing solutions. Three especially promising solution categories for the business function are (1) redefining the department's output or results, (2) restructuring connections within the department or between the department and other functions in the business, and (3) redesigning processes for managing the function.

Chapter 5 looks at the impact of information and information technology on the performance and career of the individual senior manager. We analyze the phenomenon of executive personal computing, with particular emphasis on the overlooked and subtle issues and barriers involved in adapting executive work to the potential of the technology. Using three case histories, we analyze the reasons why some executives experience little or no value from the computer and why others see it as an extraordinarily powerful force in reshaping management. We conclude that a successful executive information system requires both a vision of success and evolutionary implementation.

Chapter 6 focuses on the potential transformation of the jobs of other individuals throughout the organization. Though we briefly discuss situations in which the introduction of technology has resulted in negative transformation (such as layoffs, inordinate routinization of jobs, and excessive control and monitoring by supervisors), we focus primarily on positive transformation (opportunities for positive change in individual behavior).

In chapter 7 we address the effect of the Information Imperative on the organizational part of the business most identified with it, the information systems (I/S) function. The I/S organization is critical to a business's response to the Information Imperative because it generally delivers both technology and systems management. We begin by tracing the evolution of the I/S function through its three eras of development and by examining the dominant technologies and popular applications of each era, as well as the benefits these technologies and their applications offered, the skills they required, and the legacies they created. We summarize how four basic I/S activities (planning, systems development, operations, and technical services and support) evolved over the years. We discuss the issue of organizational ownership of the I/S function, urging that as part of the response to the Information Imperative the I/S function become smoothly, seamlessly integrated into the total operation of the company. Exactly how such integration will happen in the future remains uncertain, but we suggest two scenarios (end-user domination and the managed utility) as two extremes on a continuum of possibilities.

Finally, in chapter 8, we summarize the "shoulds" for top executives and for I/S executives as they position their organizations and themselves for transformation. We review what the corporation must do in order to transform, suggesting senior executive actions to effect changes in the industry, in the total business, in its functional units, and in individual employees. We also review what I/S executives can do in the near term to accept and manage change, as well as to position themselves and their I/S organizations for long-term success.

The Information Imperative is intended to help readers define their places in today's information age. It should serve as a lifeline, a means for gaining perspective in a confusing era. Executives and managers should gain the ability to respond to the information technology revolution, to perceive risks and opportunities, and to act in new, innovative ways. In short, they should be able to manage both their firms and their careers in an environment of enormous and continuing change.

The
Information
Imperative

1

The Information Imperative

The Information Imperative is:

> *The urgent, obligatory need for businesses (and individuals) to respond to the opportunities and competitive threats created by the revolution in information technology and by the resultant huge increase in the availability of information.*

In this chapter we will first discuss what the Imperative is not, and what it is. Then we will begin the task of addressing the Imperative. We believe that in order to meet the Information Imperative managers must first understand how it applies to their own situations. In this chapter we will illustrate its application to five different industries. In later chapters, we will consider its impacts on businesses, business functions, and individuals.

At this point the reader may wonder why, if the Information Imperative is so important, so many managers are unaware of it. Even if they are aware, why are managers unable to determine the impact of this revolution on their businesses and on their own careers so that they can respond?

One reason is that it is easy to see and measure the changes in the technology itself, but it is much more difficult to identify the impacts of new technologies on an individual's performance or on an organization's competitive position. Furthermore, information technology is only one factor causing rapid change in businesses around the world. Other well-documented economic, social, and organizational factors complicate the picture. To be understood, the information technology revolution must be understood in context.

Finally, managers are unaware or unable to respond because they don't have tools or frameworks for understanding the Information Imperative as it affects them. Accordingly, this chapter discusses the concept of the Information Imperative, and chapter 2 introduces a framework for applying it.

What the Information Imperative Is Not

Data—Or Information?

To understand the definition of the Information Imperative, it is useful to consider what the Imperative is not. It's *not* about data:

> Advances in information technology are making it easier, quicker, and cheaper to flood people with data—a lot of it useless!

Not only do technological advances inundate people with data, they also cause some people to perceive a need for even more data. As soon as the western regional sales manager for Acme Widgets has a personal computer on his desk, he will probably be tempted to request more detailed sales figures in the hope that he will be able to use this "improved" information to compete more aggressively. Or, after receiving quantities of financial data on its industry, top management may press for even more financial statistics in hopes of developing tactics to avoid a hostile takeover.

Most managers quickly learn that more data is not necessarily better information. Many clients come to us seeking ways to sort through all the data to find the few nuggets of truly important information that will make a real difference in their business performance—in particular, the nuggets that can change their decisions and actions.

It is appropriate here to distinguish terms. *Data* is the raw material; *information* is raw material processed to make it more useful; *information technology* is the means, or vehicle, used to process, transmit, manipulate, analyze, and exploit data and information. In the broadest sense, information technology encompasses all computer- and telecommunications-based capabilities currently in place or proposed for development, including databases and custom or off-the-shelf software, all components and systems that can be assembled to provide business applications.

We have chosen *not* to use the term *information systems* except in discussing the I/S function of businesses. *Information systems* means different things in different contexts: applications programs, configurations of technology, the function as a whole, and so forth.

Technology—Or Its Uses

The Information Imperative is not about raw data, nor is it about the technology per se:

> Information technology, by itself, has no intrinsic value in business.

Information technology alone does not pose a direct threat to businesses, nor can it provide revitalization and survival. For example, it is not hardware or software or telecommunications that enables Japan to design and deliver new automobile designs in less than half the time required in Detroit. Instead, value for a business lies in the *judicious use of information technology,* in prudently adapting it to fit the goals and culture of the organization and its people. Further, judicious technology uses by competitors, buyers, or suppliers can create challenges and threats for a business. Uses of technology that are not driven by business goals and are not adapted to organizational culture provide little if any competitive value.

What the Information Imperative Is

The Imperative Is about Information

Information, like capital and labor, is a resource. First, the Information Imperative focuses attention on *information* and its role in today's business environment. Information has become a central and primary resource in addressing business problems. It is the key to discovering competitive threats and opportunities. It is also critical for implementing business strategies and tactics—for deciding what to do and for monitoring, evaluating, and adjusting actions in midcourse.

Resourceful managers and analysts can piece together information from a variety of sources to identify threats and opportunities. For example, consider a manager for a U.S. auto manufacturer who sees an auto industry newsletter revealing that the Koreans are expanding the capacity of an automobile plant in Seoul by 20 percent. That news makes the manager recall a discussion with the firm's western regional sales manager, who reported that someone was approaching some of the firm's dealers, asking whether they would consider adding a new low-cost brand of cars. Our manager then collects information about Korea's auto industry, about its government policies regarding exports, and so on. By means of a personal computer the manager explores several scenarios of the financial implications of increased competition from low-priced Korean cars.

Such individual resourcefulness in finding threats or opportunities is admirable, but it is frequently not enough. Managers are realizing that they need more regular and more organized ways to capture and use information, to recognize threats, and to identify opportunities. They don't want to stifle individual resourcefulness. They want instead to support individual insights and to help perceptive individuals look in the right places at the right times, time after time.

Managers also want to use information to address broader questions: for developing business strategies and for implementing those strategies successfully. It has become fashionable to criticize U.S. businesses' lack of success in making their strategies, even promising and creative strategies, actually happen in the marketplace. We believe the criticism is appropriate. Too many strategies seem to hit an "implementation wall." One key reason is that managers do not define the measures that will indicate how the strategy is working and, even more important, that can be the basis for midcourse corrections and refinements. Further, those managers do not establish information processes to monitor and report the key measures.

For example, many companies today say that outstanding customer service is an important part of their strategies. Too few of those companies have effective information processes to support and guide their efforts. To begin with, they may not develop clear definitions of how *customers* perceive and measure service levels. Even if they have considered how customers measure service, they may not monitor those service measures. When problems arise in reaching the target for some service measure, the managers may not analyze whether it is worth while to fix the problems (however difficult); instead, it might be better for the company to find another way to meet the underlying customer need that gave rise to the measure. Finally, the companies may fail to report back regularly to customers about how they are doing on customer service.

Information is increasingly important to business success. To be valuable, the information must of course be used. Unfortunately, using information effectively is not easy. Thanks largely to the information technology revolution, executives are swamped with information from both inside and outside the business. The way to survive the information technology revolution is not to drown (or die of thirst) but to *manage* information in an aggressive way, using it to reshape the business so that it can compete in today's complex arena.

The Imperative Is Urgent and Obligatory

The second focus of the Information Imperative is on *the imperative:*

> American businesses and industry must take action. They cannot do nothing and survive; *they must change* to respond to a more competitive environment.

Leading American businesses, and other companies too, today face organizational crises, foreign competition, escalating costs, and unprecedented changes in their markets. For these firms, business as usual will no longer do. For them the prescription is: "Innovate, adjust, or die."

Ironically, information technology is both part of the problem and part of the solution. Although information technology did not create the forces of economic and competitive change, it is a key to enabling and accelerating change and the impacts of change. Information technology has also become one of the principal tools for responding to most of today's business pressures and environmental threats. (We will explore this role of information and technology in the remainder of this and several following chapters.) The potential of information as a business tool itself helps create the imperative. Some managers master the tool and use it competitively. Consequently, other managers have even more need to learn about that tool—to meet or, better yet, to anticipate its uses by competitors.

Applying the Information Imperative

In our opinion, the first step in meeting the Information Imperative is to understand the Imperative in one's own industry, business, and career. Too many managers are still missing a detailed working knowledge of what information is and what it can be. Very few understand what information technology is and how it can be employed at a practical, useful level for business (and personal) ends. In some cases, added perspective about the Information Imperative will cause managers to think in a completely new way about their businesses or careers. In others, it may merely reinforce and sharpen their current thinking. In all cases, it should help them compete more effectively.

The impact of the Information Imperative varies widely depending on the industry, business, or individual situation. We will use five industries as illustrations: newspaper publishing, automobiles, financial services, consumer packaged goods, and apparel. Of these industries, newspaper publishing and especially financial services have already been affected strongly by the Information Imperative. The apparel industry has been affected least to date. The automotive and consumer packaged goods industries lie between the extremes.

Newspapers, Automobiles, and Financial Services

Newspaper publishers have been in the information business for centuries. Relatively recently, the publishing industry has recognized the Information Imperative and its implications for the newspaper business. Instead of thinking of themselves as locked into producing a printed product with typesetting equipment and printing presses, publishers have begun to reexamine how they can use information technology to communicate information to the customer. They are looking at new ways to bring the basic product of information to its users. For example, the *Wall Street Journal* and *USA Today* have adopted distributed printing in which the content of the paper is transmitted electronically to several geographically dispersed locations, where the physical printing takes place. As a result, the papers can serve much larger markets than would have been feasible without the technology. Some publishers are exploring or experimenting with substantially different media, such as cable TV. Industry observers expect continued innovation and change. The Information Imperative has had, and will apparently continue to have, a strong impact on the newspaper publishing industry.

In the automobile industry, the Information Imperative has not yet had as strong an impact. Even so, technology is important in current attempts by domestic automakers to overcome foreign advantages in price and quality. General Motors has taken the lead and made enormous strategic investments in manufacturing technology aimed at improved quality and lower cost. It has purchased EDS, the computer applications firm. It is also building new linkages to dealers and consumers. These moves are built on information technology. By contrast, Ford has chosen to rely more heavily on subcontracting, making only specific focused investments within its own manufacturing plants; its choice demands less investment in manufacturing technology but more systems integration with suppliers.

We will look next in greater detail at selected aspects of the three other industries (financial services, consumer packaged goods, and apparel) to illustrate further the varying impacts of the Information Imperative on different industries.

Electronic Currency Exchange Has a Global Impact

Perhaps no activity illustrates the force of the Information Imperative more clearly than does currency trading. Transcontinental telecommunications networks and computer-supported trading at the major international currency exchanges enable money traders to monitor worldwide currency values continuously and to exercise buy or sell transactions essentially instantaneously. Performed in the electronic equivalent of Adam Smith's ideal market with perfect information, the traders' transactions define the real worth of a country's currency at any given moment.

The automation of currency trading represents a substantial change in speed and efficiency over former bidding processes. Most currency traders have invested in new information technology to track the "real time" value and inventory of their currency. Competitors who have not yet invested in the new technology are being forced either to operate through their competition or to provide alternate services. They cannot compete alone. To see why, imagine the plight of bank traders trying to maximize the value of a currency portfolio by using daily, batch-computer printouts as their source of information. Their efforts would be like bidding on a house by mail while a competitive buyer and the seller were negotiating in person!

While the direct results of this change in trading have been dramatic, the indirect results have also been radical. Electronic currency transfer and electronic trading have strongly affected policies of governments and have changed the standard by which currency is measured. Walter Wriston has observed that the new global trading system represents an entirely new standard for currency evaluation.[1] No longer can a central bank adhere for political reasons to a fixed exchange rate or a gold standard. The only relevant standard today is the information standard: the value determined by the free market.

This change did not require the agreement of all the banks in the financial services industry (although the introduction of industry standards made the related development of electronic currency transfer possible), nor did it require a government-sponsored treaty. It merely evolved as, one by one, international banks realized the benefits to be gained from information technology.

The Information Imperative in the financial services industry is in an advanced stage. There is probably no bank executive alive who doesn't understand that information technology is a vital component of strategy. Practically every service must be examined in terms of information transfer and information processing. Nearly every business in the industry faces a strong imperative to use information technology for interpreting and managing the information flows to and from its customers and for creating new products for the marketplace.

Scanners Give Retailers an Edge

The consumer packaged goods industry is in an intermediate stage with regard to the Information Imperative. To examine this industry, we will focus on the retailers who sell packaged goods. Most major supermarkets have begun to treat information as a valuable resource. Their growing sophistication regarding information and information technology is changing the dynamics of the packaged goods industry.

Data from scanners at the checkout counter is central to the change. Scanning technology now enables large grocery store chains to store and retrieve information on items purchased, the date, prices, and store locations. The computer can tally the sales of each item and may automatically signal the stockroom, warehouse, or supplier when restocking is required. The information gives the retailer greater control over store inventories and allows moves toward "just-in-time" stocking, saving money in shipping and storing goods. Retailers are also beginning to use the data to observe and document trends. One (hypothetical) example: the salad bar works better in the front of the store in the center city and toward the back of the store in the suburbs.

As of this writing, scanner technology has been an aid primarily to large retail grocery chains. In this traditionally low-margin business, competitors without scanners are already suffering. A few are searching for niches that may not depend on information technology. Without question, there is still a place for the "mom-and-pop" convenience stores or for the ethnic or gourmet stores common in urban areas, but how *large* a place remains to be seen.

What further opportunities does scanner technology present? Retailers appear limited merely by their own imaginations. For example, pilot projects are already underway to analyze the contents of individual shopping carts: in other words, to understand what products are purchased in combination with what other products. Such information can guide retailers in designing promotions, in issuing coupons at the point of purchase, and in other merchandising decisions. Other pilots are typing the shopper's name and address to the list of purchases at the checkout counter. Such information would make it easy to direct specific promotions to certain customers, such as those who consistently buy peanut butter or disposable diapers. The retailer might find it useful to identify and design promotions for regular shoppers (or, alternatively, for infrequent ones). The retailer might even provide a delivery service of frequently purchased staples to shoppers' homes for a small fee.

To give another example, several chains in partnership with local banks are accepting payment by bank credit card through a slot at the checkout counter. Such automatic debiting of shoppers' bank accounts has far-reaching effects. It eliminates float (and clears checks immediately), thus giving the supermarket use of the money for the days it previously took for the checks to clear. The store could provide additional banking services, charging a modest fee for a deposit or for providing the shopper's current bank balance. (Already some supermarkets have allowed banks to install automatic teller machines [ATMs] in their stores, but only after arranging to receive a fee for every customer transaction.)

How does the Information Imperative affect the relationship between the supermarket chain and its suppliers? In the past, suppliers like General Foods, Pillsbury, Kraft Foods, General Mills, and Quaker Oats obtained substantial information about the performance of their products. Some information they collected themselves. Other information they purchased from suppliers such as Nielsen (which tracks the movement of the product from the retail shelf) and SAMI (which tracks withdrawals from retail and wholesale warehouses). Part of the manufacturers' relative power was based on the inability of the retail chains to track the huge volume of items they stocked and sold. Thanks to scanner technology, the retailer now has practical access to precise data on the performance of each product in each of its stores. As a result, retailers are gaining power relative to suppliers. Imagine the interchange between a supplier's sales representative and the retail buyer today:

Supplier: Charlie, we have a new cheese that is selling very well in our test market. It is a tastier, sharper version of the Chippy Cheddar that has done so well for you. And we're offering it in two sizes—the big one for families and the small one for single-person households.

Buyer (turning to the terminal): Sounds interesting, Margaret. Let's see. Hum . . . Chippy Cheddar . . . Well, sales are off in our affluent-market stores this week. Down by 30 percent! Holding constant in the factory district, though. It will be hard for me to justify stocking two sizes. When we dropped one size of your herb and cheese spread as an experiment in a few stores, we didn't see much impact on total sales dollars.

Supplier (straining to see the buyer's screen): Gee, you didn't?

Scanner technology, then, has serious implications not only for the retail grocery chains, but also for the major suppliers. Manufacturers see their slow-moving products occupying less and less shelf space. They find retailers far less willing to accept their advice and guidance, and may find it necessary to buy the retailers' scanner data.

The time is imminent when managing information *about* packaged goods will be more important competitively than the physical handling of the products. Information and information technology will continue to bring further change. The retailers may, for example, ultimately have enough information to eliminate the need for warehouse buffer inventories. Suppliers envision increasing pressures for frequent, smaller deliveries, and they fear increased distribution costs. Both retailers and manufacturers will need even better consumer information in order to prosper competitively.

Perhaps the supermarket of the future will not be a physical location at which consumers shop but will instead be a very sophisticated broker between the consumer at home and the grocery manufacturers. Such "retailers" might accept orders by voice phone or computer link. They might themselves assemble the orders and deliver them to consumers, or they might instead pass the orders to a separate fulfillment operation. On the other hand, perhaps consumers will insist on shopping in a store. They may value both the entertainment value and the potential cost savings of active shopping. In that case, we can expect to see technologically sophisticated stores using information to tailor their offerings for their particular target markets and to operate efficiently.

To summarize: information technology is bringing more and better information to the retail grocery chains. As a result, these chains are gaining power and economic leverage relative to manufacturers. It is imperative that both retailers and manufacturers take notice and, most important, that they find effective actions to deal with the changes. If the time for procrastination in facing the Information Imperative is past in currency trading, it is critically short in food retailing and manufacturing.

The Apparel Industry Faces the Imperative

The U.S. apparel industry vividly illustrates the problem of vulnerability to foreign competition. U.S. clothing manufacturers have been steadily losing market share to offshore manufacturers. They have been unable to compete successfully against the labor cost advantage that foreign (mostly Asian) producers enjoy. Some industry watchers claim that U.S. manufacturers cannot overcome that disadvantage.

Other more optimistic observers emphasize the U.S. industry's potential advantage: its geographical proximity to the markets. So far, at least, it is still prohibitively expensive for the importers to fly goods to the United States. Domestic producers might be able to exploit this fact if they could shorten their own production-to-consumer cycle.

How might they do so? A 1985 *Wall Street Journal* article described what it called an eleventh-hour effort by U.S. textile and apparel makers to become competitive with imports.[2] The article explains that the buying and production cycle for apparel often takes well over a year (approximately sixty-six weeks). Production commitments must be made long before the consumer sees a garment in a store. Unwise buying decisions are disappointing for all. If the retailers order too much (the customer doesn't buy as expected), the retailer is stuck with slow-moving merchandise. On the other hand, long buying cycles mean that retailers cannot restock if they order too little; they simply run out. If manufacturers could get feedback from retailer customers sooner, they might be able to reduce both types of problems. They might be able to get additional merchandise to the stores on items that sold well; they might also allow smaller initial orders and thus help avoid overstocking items that didn't sell well.

Such a dramatic change is feasible, if at all, because of information technology. The department or clothing store could capture information, purchase-by-purchase, at the cash register. The information could help the store's buyers and the manufacturers too. The article quotes one participant in the effort who estimated that the buying and production cycle could be cut (from sixty-six) to twenty-one weeks.

To make such a business system work, with today's basic ownership structure in the industry, the major segments of the apparel industry would have to cooperate and transform their businesses. Textile mill producers in South Carolina, clothing manufacturers in New York, intermediate distributors, and retail stores would have to form unprecedented liaisons. Firms that had heretofore acted as separate entities would have to function as a coordinated production and distribution system.

In short, the industry would have to reexamine the long chain of events—from purchase all the way back to the manufacture of fiber—in order to develop a new, shorter cycle that responds more quickly and accurately to the demands of the marketplace. Information technology integrated with business procedures would hold the industry chain together in the face of foreign competition.

The optimists may or may not be right about this or other possible scenarios for the future of the U.S. apparel industry. Regardless of whether they are right, this case provides an example of an industry in which the Information Imperative seems not to have arrived yet. Even in such industries, however, the Imperative is coming. We believe that in most industries where it is not yet felt it will arrive soon, in force.

Responding Successfully to the Imperative

Success Requires Transformation

These industry illustrations of the Information Imperative suggest three points:

1. The Information Imperative is different in each industry.

2. Each industry is vulnerable to the effects of the Information Imperative.

3. Businesses must generally undergo substantial change—or transformation—to take advantage of the opportunities that the Information Imperative unleashes.

The five preceding examples should verify the first point. The Information Imperative is hitting different industries at different times and at different rates. Financial services has been immersed in it for several years, retail food is in its midst, and apparel has not yet been strongly affected. The Information Imperative is inevitable but will reach different industries at different times.

All industries are vulnerable in part because the potential of information and technology is so great: sooner or later some competitor will use them effectively. Industries are also vulnerable because their supplier industries and customer industries are vulnerable and will change. In a sense, the Information Imperative creates a domino effect.

These examples also demonstrate that much more than information technology needs to be managed. Technology will help, but for most businesses true success requires substantial adaptation, or sometimes changes in the fundamental nature of the business—what we call *transformation*.

Changes affect individuals, too. When one fifty-year-old automotive engineer started his career in the 1950s, engineers began the design process for new automobile bodies by making temporary clay models. When they were satisfied with the design, they made permanent mahogany models. From these they derived the aluminum templates that they sent to a die maker, who built the stamping machines for the new automobile bodies. Today, automotive engineers still build clay models, but then they scan those models with cameras and transmit the information digitally into computers where they can manipulate and change the models as they watch on CRT screens. When the engineers are satisfied, they transmit the information electronically to the die maker who builds the parts. As our fifty-year-old engineer pointed out, "In our business today, engineering *is* information." For him, the evolution from physical models to electronic ones has been a crucial transformation over his career.

The examples in the chapter illustrate the Information Imperative:

> In individual careers as well as in industries and in businesses, the focus of attention is shifting from physical products alone to the product and information as well, from old ways of thinking about business to new ways, from old forms of information transfer to new forms. All of these changes are mediated or supported by information technology.

Putting on the Information Lens

To reiterate: The Information Imperative is the urgent, obligatory need for businesses (and individuals) to respond to the opportunities and competitive threats created by the revolution in information technology and by the resultant huge increase in the availability of information.

To be successful in responding to the Imperative, managers must first understand how it applies to their own situations. In essence, they need to view their surroundings through an "information lens." Just as a "financial lens" allows managers to understand the meaning of raw financial data, so the information lens enables them to focus on the risks and opportunities resulting from or related to the revolution in information and in its technology.

The examples of this chapter illustrate the use of the information lens for viewing entire industries; chapters 3 through 6 demonstrate its use for businesses, business functions, and individuals. First, however, chapter 2 introduces another important tool for meeting the Information Imperative. It presents a framework that helps in understanding the business value of current applications of information technology and in identifying promising future applications. That framework helps give structure and meaning to data coming through the information lens.

Notes

1. W. Wriston, "In Search of a Money Standard: We Have One: It Comes in a Tube," *Wall Street Journal,* November 12, 1985, p. 28.
2. D. R. Sease, "Move to Fight Apparel Imports Is Set: Textile, Garment, Retail Concerns to Join Forces," *Wall Street Journal,* December 17, 1985, p. 6.

2
Benefits and Beneficiaries

B efore they can respond to the Information Imperative, managers must translate the concept into terms that are meaningful in their own environments. Many managers have difficulty making such a translation.

One common problem is focusing too much attention on the
technology itself. When Company A gives personal computers
to its sales force, the sales manager of competing Company B
often asks, "Should we also give personal computers to our sales
force?" The questions that *should* be asked are: "What is the
competitor's business objective? What should be our business
objective? How does technology fit?" Or, after reading dozens
of articles in the business and trade press about how information
technology is helping corporations compete more aggressively,
an executive might ask, "What can computers do for us?"

A "technology first" approach seems to support the view of
Joseph Weizenbaum of MIT that technology is "a solution in
search of problems."[1] The business scene is littered with expensive systems authorized by managers who, inexperienced and
influenced by computer mania, invested in technology without
thinking through their business needs first.

Although an awareness of technology's capabilities is useful
and necessary, managers' basic response to the Information Imperative should be to look for ways to improve or even to transform their businesses. Rather than asking about specific kinds

of computers or information systems, managers should ask the following basic questions:

> What are our business goals? How can we use information and information technology to achieve our goals?

> What new business goals should we establish? Does technology offer us possible new goals?

> What changes in the strategy, structure, systems of management, and work behavior will be needed in this firm to meet our goals? How can applications of information technology help bring about the needed changes?

The key to success is to start by focusing on the business and work change. Decisions about information technology can then follow. Once stated, the principle seems simple and indeed obvious: business needs first, technology second. Why then is this principle not applied routinely in practice?

We believe one reason is that managers do not stop to think about the principle of business needs first. Information technology seems foreign and esoteric; many managers assume that it should be managed in a separate world, as a separate function.

A second reason is more challenging. Even when managers want to drive technology from business needs, they generally do not know how to do so. They lack the processes and frameworks for connecting technology choices with business needs. In this chapter we present a framework or tool that can overcome this common problem and enable decision makers to respond to the Information Imperative. We will apply the framework in the remainder of the book.

A Framework for Managers

We call the framework the Benefit/Beneficiary Matrix. It focuses on the type of value to be gained from an application of technology and the organizational unit employing the application. The Benefit/Beneficiary Matrix provides a diagnostic tool for executives who wish to assess where their organization and its parts are with regard to information and information technology. It also provides a tool for planning and communication, for determining what the organization's technology efforts should be, and for obtaining understanding and commitment for decisions regarding information technology.

The origin of the matrix helps explain the philosophy behind it. In 1984 the Index Group and Hammer and Company completed a joint study entitled *Managing Personal Computing*. The study began with the purpose of exploring only personal computers and their management. The researchers concluded that business executives "need a conceptual framework for understanding personal computing in the context of all computing."[2] They found that both business managers and information systems groups were guilty of focusing on the technology rather than on the business benefit. They then designed and introduced the Benefit/Beneficiary Matrix to enable managers to focus on and define the business *benefits* of information technology and the target individuals or group—the *beneficiaries*—who would obtain the benefit. They believed that the value of the Benefit/Beneficiary Matrix "is that it is independent of any particular current or future form of computing technology."[3]

The Benefits of Information and Technology

The Matrix (figure 2–1) identifies three types of business benefits from applications of information technology:

1. Efficiency

2. Effectiveness

3. Transformation

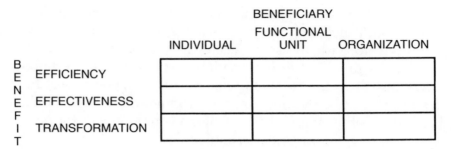

Figure 2–1. *The Benefit/Beneficiary Matrix*

	BENEFICIARY		
	INDIVIDUAL	FUNCTIONAL UNIT	ORGANIZATION
B E N E F I T EFFICIENCY			
EFFECTIVENESS			
TRANSFORMATION			

Figure 2–2. *Efficiency*

Applications in the *efficiency* category (see figure 2–2) allow users to work faster and often at a measurably lower cost. Automating the clerical work in payroll or accounts receivable, using computers to handle paperwork in issuing insurance policies, or automating the generation of standard but personalized (form) letters are primarily efficiency moves.

Applications in the *effectiveness* category (see figure 2–3) allow users to work better and often to produce higher-quality work. For example, an automated accounting system may go beyond routine calculations and record keeping and may in addition provide carefully constructed summary reports that help managers make better decisions. Automating the preparation of purchase orders may be primarily an efficiency move, but users can proceed to the effectiveness category if they capture the in-

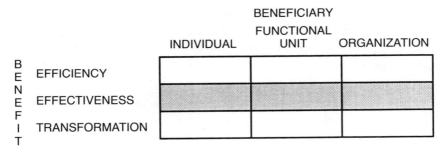

Figure 2–3. *Effectiveness*

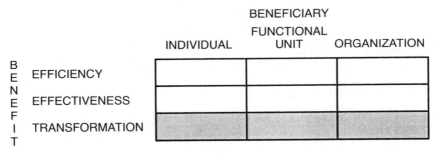

Figure 2–4. *Transformation*

formation about purchases and use it to negotiate volume discounts or other concessions from their suppliers.

Applications in the *transformation* category (see figure 2–4) change the basic ways that people and departments work and even change the very nature of the business enterprise itself. For example, consider the purchasing department of a research-oriented organization that buys substantial amounts of expensive equipment. Efficiency moves help the department generate more purchase orders faster. Effectiveness moves help it negotiate discounts, monitor the performances of its suppliers, and so on. More important, however, the information the purchasing department accumulates can also be used to transform that department's role. Suppose it keeps track of where in the organization instruments go and what their projected (and actual) usage levels are. Then, if a manager requests the purchase of another specialized instrument, the purchasing department can first scan its record of the current instrument inventory to determine whether the manager's needs could be served with spare time on an instrument the organization already owns. Such activity provides efficiency and effectiveness, to be sure—but it also transforms the role of the purchasing department from pure purchasing to resource management.

Other broader examples of transformation include some of the most impressive uses of information technology for competitive advantage. American Airlines' famous SABRE reservation system fundamentally changed the relationships among the airline, travel agents, and others in its industry. Some automated systems for entry (like American Hospital Supply's famous ASAP system, which is discussed in the next chapter) are also examples of transforming moves.

The Beneficiaries of Information and Technology

The Benefit/Beneficiary Matrix also identifies three categories of beneficiaries for applications of information technology:

1. Individuals

2. Functional units (or business departments)

3. Organizations

Many applications, including the great majority of early applications, primarily benefited the functional unit or department (figure 2–5). Payroll systems made the payroll process faster and more accurate; their main impact was in the accounting department. Similarly, automated accounts payable and general ledger systems have the accounting department as their main beneficiary. When inventory control was automated, the results were felt primarily within the production department in most companies.

More recently, information technology applications have be-

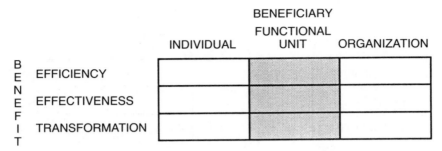

Figure 2–5. *The Functional Unit as Beneficiary*

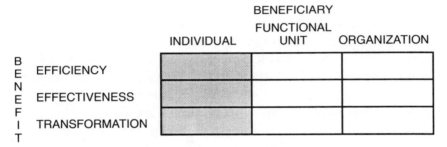

Figure 2–6. *The Individual as Beneficiary*

gun to address the needs of individuals (see figure 2–6). Many but not all such applications are used on personal computers. For example, consider a brand manager in a consumer products company. In the past, she likely received quantitative analyses of sales and related variables from her brand assistant and perhaps from the marketing research department. With the advent of personal computing she has augmented those analyses with spreadsheet calculations on her own computer. The additional calculations are not intended for others; they help the brand manager understand and explore the quantitative results in her own terms.

Some of today's most powerful and important applications benefit the entire organization (see figure 2–7). Some link a company with its suppliers or with its customers; American Hospital

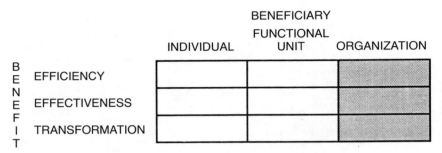

Figure 2–7. *The Organization as Beneficiary*

Supply's automated ordering system, ASAP, is a good example. Merrill Lynch used information technology to create its cash management account product, tying together several financial capabilities (such as stock purchase and sale and checking accounts) to serve its customers and change Merrill's relationships with them. American Airlines' SABRE system is a third example. Applications of information technology can also benefit an organization if their primary impact is to tie together the various departments of a company and to help them work better together. An application integrating order entry, sales forecasting, production planning, pricing, and other functions might provide important benefit to a company (and allow it to serve its customers better too).

The Framework Provides Historical Context

In the next section we will consider the most important uses of the Benefit/Beneficiary Matrix, which are in planning and management. First, however, we will use the framework to describe the evolution of applications of information technology in business. Our brief historical side trip has two purposes. First, historical examples provide illustrations for the reader of how applications fit into various parts of the Matrix. Second and more important, the history of applications has created some of the

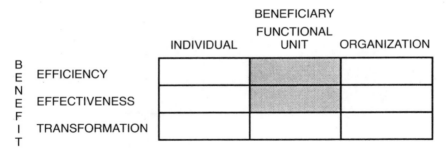

Figure 2–8. *Domain I*

barriers to success in meeting today's Information Imperative. We will briefly consider the history here to provide context. We will return to a more complete consideration of the history in chapter 7's discussion of the information systems function.

Most companies started to use computers in the late 1960s or even later; only a few pioneers began their computer histories in the 1950s. From 1960 to the late 1970s, computers served as little more than fast calculators designed to facilitate cost cutting in existing functional units such as accounting, purchasing, and manufacturing. The primary benefit from these early computer applications was increased efficiency—the ability to perform existing operations faster. As firms grew adept at efficiency applications, subtle gains in effectiveness began to appear. Business functions learned to perform their work better and attained higher standards of accuracy and quality along with greater speed. Information from these early applications began to help functional managers identify potential problems and thus make better decisions. The first domain of the Matrix (see figure 2–8) represents this early period in which traditional, "back office" routine or clerical applications improved efficiency and effectiveness within functional units.

With the advent of microprocessor technology came personal computers and the end-user computing phenomenon. Individual users came to depend on spreadsheet software on their personal computers to do their work faster and better. As man-

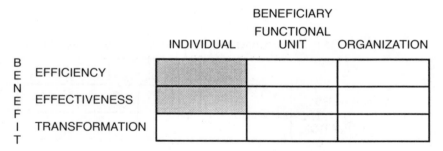

Figure 2–9. *Domain II*

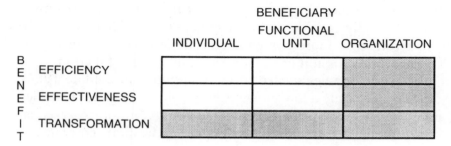

Figure 2–10. *Domain III*

agers gained access to more and better information, they began to perceive technology as a benefit to their own productivity and performance. Thus the second major period of computing history is represented by the second domain of the Matrix (see figure 2–9), where end-user computing applications improved efficiency and effectiveness primarily for individuals.

Today, information technology can deliver the business benefits of efficiency and effectiveness, not just to functions and individuals, but also to the organization as a whole (see figure 2–10). In addition, information technology applications can bring about basic changes or transformation—in the jobs of individuals, in the roles of business functions, or in the organization as a whole. Domain III consists of all applications that focus on the organization as a whole and of all applications whose prime benefit is transformation.

Domain III information technology applications are usually especially challenging ones. They are particularly likely to involve a mix of information technologies—communications and other technologies as well as computers. In addition, while the progression from the first to the second domain was a relatively natural evolution, the progression from the second to the third domain represents a giant leap forward. Transformation benefits, particularly for the organization as a whole, are much more than large amounts of efficiency and effectiveness. They represent something new and different. Domain III applications can be substantially more powerful in their impacts than are applications in Domains I and II.

To date, there have been relatively few deliberate implementations of applications to transform organizations; most of those that do exist have resulted from chance discoveries or have evolved from applications developed in earlier eras. But organizations today are seriously and more systematically looking for opportunities within Domain III; it is just a matter of time (and not very much time) before many businesses begin to employ technology more aggressively for their own competitive advantage. Domain III information technology applications will be particularly important in enabling businesses to respond to the Information Imperative.

Using the Matrix

The Matrix as a Planning Tool

The most important use of the Benefit/Beneficiary Matrix is as a planning tool. The Matrix helps managers understand and describe their current portfolios of information applications. It also helps them to describe desired future application portfolios and to manage the changes in their portfolios over time.

Step 1: Taking Stock. The first step in applying the matrix as a planning tool is to take stock of the current situation. The following exhibit, "Taking Stock of Current Applications," lists the key parts of this task. Both information systems (I/S) managers and other business managers should participate. In order to focus on the business value of the applications, the business managers should take the lead in identifying the primary benefit and the primary beneficiary for each application. Note that the process includes all important information applications, both those that are currently implemented through technology and those that are not.

Step 2: Evaluating the Current Mix. Most managers will find that the overwhelming majority of their current applications are in Domain I and that most emphasize efficiency benefits (rather than effectiveness) even within that first domain. Examples are automated accounting systems and some simple inventory and manufacturing applications. This emphasis on Domain I efficiency is not surprising given the history we have just sketched for applications. And to be sure, Domain I applications are often extremely useful. Because of the power of the information technology revolution and the urgency of the Information Imperative, however, most managers will decide that Domain I applications are not enough.

Many managers will find some Domain II applications in their current portfolios too. Spreadsheets and word-processing applications for the individual are especially common Domain II applications. Again, they can be extremely useful. And again, they will generally not be enough to meet the challenges of the Information Imperative.

Finally, managers may find some Domain III applications even in their current portfolios. They may also find existing applications that could be used in more powerful ways, providing some Domain III benefits. The inventory should note such applications and such opportunities.

Taking Stock of Current Applications

1. Enlist both information systems (I/S) professionals and other (business) managers in the process.

2. Have the I/S professionals inventory all current information systems—that is, the technology-based applications.

3. Have the business managers inventory all other applications of information—that is, those that are not currently implemented through technology.

4. For each current application of information, or of information and technology, identify the *primary* benefit:[a]

   - Efficiency (doing the same things faster or less expensively)
   - Effectiveness (doing things better)
   - Transformation (doing fundamentally different things)

5. For each current application, identify the *primary* beneficiary:[a]

   - The individual
   - The functional unit or department
   - The organization as a whole

[a]There will be one *primary* benefit and one *primary* beneficiary for most applications. It is acceptable, however, to identify a few applications as having two equally important benefits (or two equally important beneficiaries).

Identifying the Target Mix

1. Have business managers list the new applications they consider most important, giving the primary benefit and the primary beneficiary for each.

2. Have I/S professionals and business managers jointly identify the existing applications that need maintenance or replacement and that they believe provide enough business value to warrant the effort. They should specify benefit and beneficiary for each.

3. Search for important new ideas for applications: through reading and education, by looking at other companies in many industries for ideas, through discussions.

4. Using decision-making processes that fit the particular company, outline the mix of applications that seems to fit its situation and future needs.

Step 3: Identifying the Target Mix. After taking stock and evaluating the current mix, the third step is to decide on the approximate desired future mix of the three domains. Choosing that desired mix is *not* a job for the I/S manager alone. It is a critically important strategic business decision and should involve (and usually be driven by) other very senior managers, although the chief I/S manager would normally participate. The exhibit "Identifying the Target Mix" outlines this challenging step. Inventorying known opportunities for new applications is the easier part of this step. The harder part is to create additional ideas. Education and reading help; for example, this book contains examples that will suggest opportunities to some readers. We suggest looking at a wide range of industries for ideas about applying information technology. Some basic education about the capabilities of technologies helps. Most important, managers should view the world through the information lens, which was introduced at the end of chapter 1. They should work to *see* as many as they can of the business applications of technology that surround them. They should work to *understand* the business importance of those applications and then to consider whether analogous applications could bring competitive advantage in their own situations.

As they proceed to outline a desired mix of applications, most managers will want to include applications in all three domains. Domain I remains useful and important; most companies will find new applications for functional efficiency and effectiveness, and they will also have to perform maintenance and replacement of key existing Domain I applications. For example, managers in one company might identify a high-priority new application for improving sales deployment (an *effectiveness* move for the sales *function*). They might also decide that key Domain I applications for the finance *function* merit replacement with more *efficient and effective* versions.

Similarly, managers in most companies will want to continue to reap the efficiency and effectiveness benefits for individuals in Domain II. Many will rely on individuals to do the bulk of the work in Domain II; the company may need to provide only support and advice.

Most managers will identify a key objective of increasing applications in Domain III. The opportunities they identify will depend on their current mix and especially on their business strategy. A company with the strategy of being the low-cost producer in its industry will likely value efficiency and effectiveness moves for the organization (as well as Domain I moves for its functions); however, even this kind of organization should not ignore possible transformations that might, for example, change the basic cost structure. A company competing on the basis of differentiation and service to its customers would not entirely ignore efficiency, but its managers will usually want to emphasize the benefits of effectiveness and transformation for the organization and its departments and individuals.

Most managers will feel that they cannot identify the right mix easily or precisely. For one thing, they and their colleagues will usually lack a thorough understanding of Domain III applications, and they will often have trouble defining such opportunities. Nevertheless, the Matrix remains useful. For instance, the desired portfolio can serve as a flexible goal and as the basis of discussion. Indeed, it can serve as an impetus to managers to generate the ideas that will change the mix to make it more like the goal.

One of our clients provides an illustration. In that company, functional unit managers submitted new applications proposals to top management for approval and funding. When the Matrix was employed for the first time, some proposals, particularly functional-efficiency applications, were rejected because they were too expensive for the type of benefit provided. The managers who submitted those proposals were encouraged to add capabilities to their proposals that would expand the benefits, at least to improve effectiveness but perhaps also to create a transformation application. Many proposals emphasizing benefits for individuals were also sent back for further development with the challenge issued to the managers of expanding them at least to the functional unit level. Sound proposals that fell within Domain III were funded immediately, and their sponsors were recognized and encouraged to continue to identify such opportunities in the future.

This example shows how the Matrix can help to generate ideas and to explain decisions about the allocation of resources for implementing information technology applications. The Matrix can then be used repeatedly over time for mapping the current portfolio, for describing the desired portfolio mix, and for communication and discussion. Thus, it can help in monitoring and managing a company's applications of information and information technology.

The Matrix as an I/S Management Tool

> The Benefit/Beneficiary Matrix can also help I/S professionals to manage more effectively.

Information systems managers can be among those most guilty of interpreting and responding to the Information Imperative in purely technical terms. And too often they try to manage information systems on the basis of the technology these systems use rather than on the basis of the business purpose they serve. For example, many I/S managers virtually ignore personal computer–based applications, assuming they require a minimal level of control and management attention. Conversely, these managers often exercise tight control over the applications that run on the central computer, because such applications are easier to find and because they have been "doing it that way for years." The problem with this hardware-oriented approach to I/S management is that some personal computer applications are extremely critical to the business, while some mainframe computer applications are routine, low-risk systems. The first often require considerably more careful management than the second.

The Benefit/Beneficiary Matrix suggests instead an applications approach to I/S management. By placing each application in the appropriate cell of the Matrix, an I/S manager can create an overall portrait of the applications in the organization and can develop guidelines and management controls for each area. The individual efficiency and individual effectiveness cells usually require little attention from senior management or from the I/S function; instead, their development, usage, documentation, audit, and maintenance could be largely the individual user's responsibility. On the other hand, any Domain III application involving customers or suppliers would require senior management attention *no matter what the technology.* The I/S function might be responsible for the development and supervision of such applications, but senior line management would define and approve them and would enforce appropriate controls. In general, Domain III applications might require corporate- or divi-sion- level approval; departmental or user approval might suffice for Domains I and II.

Keys to Domain III Applications

One basic key to successful Domain III applications is the point we have been stressing throughout these early chapters: *Drive the applications of information and information technology from the business needs, not vice versa.* This statement holds true for all types of applications; it is most important and most challenging in Domain III. A second key to success in Domain III is awareness and education about how information technol-ogy can contribute today to business objectives; the following chapters help by describing in considerable detail the roles of the Information Imperative and of information technology in relation to the organization, the business function, and the individual.

The third key to success in Domain III is to recognize im-portant basic differences between applications in that domain and others. In fact, the three domains of the Benefit/Beneficiary Matrix differ in terms of: (1) appropriate roles for senior man-agers and users and for I/S, (2) appropriate methods for justify-ing investments in applications, and (3) implementation method-ologies (see the exhibit "Keys to Domain III Applications"). We next consider each of these three differentiators in turn, empha-sizing the appropriate choices for Domain III.

Appropriate Roles

> Championship by senior managers and active and ongoing participation by business managers is needed in Domain III.

Domain III applications have great business importance. They may require large systems of information technologies; they almost always have large and far-reaching impacts. Appli-

---

Keys to Domain III Applications

1. Drive the applications of information and information technology from the business needs, not vice versa.

2. Actively work to learn about as many as possible of information technology's contributions to business objectives.

3. Recognize the important basic differences between applications in Domain III and those in the other domains. Domain III applications generally require:

   • Championship and participation of business managers, including senior managers

   • New justification methods

   • New implementation methods

---

cations for the organization as a whole cut across departmental boundaries (as, in fact, do some other Domain III applications), and they require coordination and cooperation from different parts of the business.

Selecting appropriate Domain III applications frequently requires considerable insight into the company's current and future strategic direction. We find that in most companies for all of these reasons, senior management championship is needed for Domain III. Moreover, successful identification and implementation of such applications requires ongoing active participation of business managers, as will be emphasized again later in the section on implementation in Domain III.

Cost Justification Methods

The cost justification methods derived (and used successfully for years) for Domain I do not fit the other domains. They can be especially disfunctional in Domain III.

During the 1960s and 1970s businesses devised control methods and budgeting methods appropriate to the I/S function of the time. Especially because technology seemed foreign to them, non-I/S managers wanted firm, understandable controls for I/S. Most applications in those days were primarily efficiency moves: they reduced headcount or provided similar quantifiable benefits. As a result, the common methods for justifying investments in information technology were extremely quantitative and often rather short term. We call such methods *"headcount and keystroke"* justification because they emphasize benefits such as lower headcounts and fewer keystrokes (and other dollar impacts of changes).

Headcount and keystroke justification often works quite well for Domain I efficiency moves, and it is entrenched in the thinking about information technology applications of many managers. The problem is that headcount and keystroke does not fit well in the other domains (or even for effectiveness moves in Domain I); it can be disastrous in Domain III. Strategic, transforming business moves are designed to strengthen relationships with customers, change the nature of the offering in the marketplace, and make other extremely important changes that cannot be reduced to simple measures like headcount and keystroke; it is especially difficult to describe their immediate impacts.

Moreover, senior managers often select such moves without fully understanding their implications; indeed, the very power and importance of Domain III applications often mean that we cannot know at the outset how they will play out. Managers select them on the basis of a combination of hard analysis, vision, and gut feel.

We find a surprising number of companies still trying to justify applications that we would classify in Domain III with the old headcount and keystroke methods. Sometimes such applications squeeze through the justification filter (sometimes with some help from creative accounting), but even in such cases the justification procedure fails to study and document the true wider business purpose for the application. In other cases, outmoded justification methods reject important and promising applications, keeping companies from meeting the Information Imperative with Domain III information applications.

Warren McFarlan of Harvard Business School tells of the I/S executive at a major bank who wanted to invest in a global network; this executive thought that the resulting exchange of information would provide benefits that he could neither quantify nor fully predict. In spite of that fact, he insisted that the network was a good investment because it had the potential to transform the organization as a whole. Top management was reluctant to approve the investment until the I/S manager justified the cost of at least one piece of the network on efficiency grounds. After considerable work, he was able to show a 14 percent return before taxes—just barely enough to get it approved. The actual installation provided considerably more hard benefits than the executive originally estimated, but the question of such benefits lost importance. Once the network was installed, top management immediately began to recognize its organizational benefits. The cost justification and cost overruns were no longer an issue; the wider business benefits became clear.

Even when the short-term numbers from a potential Domain III application cannot be manipulated to meet headcount and keystroke types of justification, those applications can be strategically powerful and extremely worth while. The leap of faith needed to implement such an application, even if it fails a rigid quantitative test, must often be made by senior management. Hence, the problems of justification methods for Domain III reinforces the point made earlier about the need for championship by senior managers.

Implementation Methodologies

Domain III applications often require new systems development methods (the key being what we call *prototyping*)

for successful implementation. Success also requires identification and careful management of the range of changes,
including organizational changes, that such applications
create.

Traditional methods for implementing information technology applications emphasize structure, planning, and control.
They start with detailed systems designs. Implementation pro-
ceeds according to a detailed and usually firm schedule. Such
procedures were developed in part as responses to typical problems in implementing early applications. Such early systems were
notoriously late and overbudget—and too often were not what
the user really wanted. I/S professionals adopted careful methods
of project management and control to overcome the early problems and to improve relations with their users.

Today's problem is that useful as it often is in Domain I, the
traditional methodology is frequently highly inappropriate in
Domain III. In a typical strategic Domain III application, neither
the users nor the I/S professionals know at the outset just what
the system should be or how it will work. A formal methodology
that begins with a detailed system design often simply does not
fit. So instead we often advocate the use of "prototyping," or
learning by doing. The users and I/S essentially begin a journey
together. They try something (generally something small) and
see how it works. They learn from the experience and improve
the initial design. In some cases, they throw out the first implementation and build a better one. They continue to experiment,
learn, and adapt.

Prototyping requires new and different skills from I/S professionals. (Some may not be able to develop those skills.) It requires active and ongoing involvement from key users. It also
requires a change in the ways both users and I/S managers view
systems efforts. We can think of traditional information systems
as built of concrete and intended to last. In that traditional systems world, programmers would likely be fired for building a

system that was thrown out almost immediately. In the proto-typing world, by contrast, we can think of early versions as built of chewing gum and baling wire. Users and I/S managers assume that early versions will be discarded—and that concrete is appropriate only for later versions that will fit the business need well and will be used for extended periods. The change from the traditional methodology is profound—but it is required for some Domain III efforts.

By their very nature, many Domain III applications create substantial changes, such as redefinitions of the roles of individuals in a company and major shifts in the relations among the functions of the business. Such substantial changes are not easy to implement. In fact, the organizational and behavioral implications of Domain III applications are often very difficult to foresee—and obviously it's hard to manage the changes if we don't even know what they will be. Nevertheless, success requires managing those changes and managing them well. It requires active consideration of the full range of changes that result from a strategic new information application, not just management of the technology itself and its most obvious changes. This need is yet another reason for the active and ongoing participation of key users and senior managers in successful Domain III efforts.

Success with the Matrix

The Benefit/Beneficiary Matrix is a powerful framework. Its use can help managers meet the Information Imperative, but by no means does it make the task of meeting that challenge an easy one. The Matrix and the keys to its use, described in this chapter, help. The next four chapters explore the implications of the Imperative for organizations, for functions, and for individuals. They contain varied examples of imaginative and important uses of information and information technology, in part to help prompt thinking about new applications. They also elaborate on the points about implementation that were introduced in this chapter.

Notes

1. Index Group, Inc., and Hammer and Company, Partnership for Research in Informations Systems Management, *Managing Personal Computing* (Cambridge, Mass., 1984).
2. Ibid.
3. A description of the Matrix can be found in C. F. Gibson and M. Hammer, "Now That the Dust Has Settled: A Clear View of the Terrain," *Indications* 2:5 (Cambridge, Mass., July 1985).

3
Threat and Response for the Total Business

Overall, the Information Imperative is an industry phenomenon which, we have argued, will sooner or later impose the necessity for Domain III applications of information technology in all businesses. As a result, all managers, senior line managers in particular, need to understand and act on threats and opportunities from technology in relation to their strategic goals and competitive moves. In this chapter we will first examine key threats facing companies in their basic tasks in the marketplace—the identification and satisfaction of customers' needs, or the *marketing task*, in its broadest sense. We will then consider how information and information technology are important contributors to the solutions to those threats—to efficiency moves, to actions aimed at effectiveness, and, most important, to true transformation for the organization and its components.

American Hospital Supply: Information Technology Helps to Transform an Industry

Many managers know that American Hospital Supply Corporation (AHSC) is a prime example—perhaps the classic example to date—of a business that has used information technology for competitive advantage. AHSC is a large distributor of hospital supplies, handling an extremely wide range of supplies from syringes and hospital gowns to simple instruments. The well-known AHSC Analytical Systems Automated Purchasing (or ASAP), the firm's computerized order entry system, enabled hospital customers to use technology in ordering supplies directly from AHSC.

While many managers know about ASAP, few recognize that the story highlights the variety of technology-related forces operating on today's businesses. To see that point, we will take a somewhat unusual perspective and will explore the impact of the ASAP system on a large, respected manufacturer of chemical supplies for hospitals. The chronicle begins long before this chemical supplier was aware of anything unusual at AHSC. Like many other suppliers, the manufacturer preferred to avoid distributors like AHSC and to sell its products directly to hospitals. The rationale, in part, was that selling specialty chemicals required a sales force more sophisticated than the AHSC sales force, who required little specialized knowledge to sell thousands of common hospital supplies, such as disposable surgical gowns.

In 1976 American Hospital Supply introduced ASAP to simplify the ordering of the thousands of individual supply items its hospital customers required. With ASAP, customers could use a touchtone phone to enter required product numbers, or they could use a wand to read bar codes from cartons or from the catalogue.

At first, ASAP seemed like a simple efficiency move, and it might have turned out to be only an efficiency move if AHSC's distributor competitors had been ready to follow suit. They were not ready, however; they had not yet computerized their own inventories and therefore could not offer automated purchasing. Some tried to catch up, but AHSC had taken the lead.

AHSC's next system, called ASAP 2, looked like yet another move for efficiency with some effectiveness benefits added. AHSC added teletypes so that customers could obtain printed records of their orders. Again, AHSC's competitors could have done the same thing, but they were not yet technically ready.

Our chemical supplier did not even notice ASAP 1 or ASAP

2. Only after the introduction of ASAP 3 did its managers become vaguely aware of the ASAP system. One of the first hints came in a strategic planning meeting when the sales vice president mentioned that it was getting more difficult to sell nonproprietary products. He linked the reason to ASAP 3. AHSC's latest system allowed customers to create single files (such as standing order files or repetitive order files) that bound together groups of common items. For example, a hospital's pathology lab might use a standing order file to handle its regular orders. That way, the customer could order an entire assortment of supplies with one file name; there was no need to enter the individual items. Such files bound individual items into groups for purchasing. Our supplier's sales representatives began complaining that customers were unwilling to break those bonds (and lose the efficiency and effectiveness benefits) to buy one or two products from them—even when those products were less expensive or otherwise preferable.

By the mid-1970s, the health care industry was already under pressure to control costs, and many hospitals welcomed the savings in ordering and inventory costs that ASAP provided. The efficiency and effectiveness benefits helped both AHSC and its customers, but the benefits did not stop there. ASAP 3 also brought transformation—a fundamental change in the way the hospitals interacted with their suppliers—and that change posed a real stumbling block for competitors.

By late 1983 when AHSC introduced ASAP 4, a computer-to-computer link between AHSC and the hospital's materials management software, many suppliers had started to pay attention. With ASAP 4, ordering was automatic (though subject to human override). There was room for only one distributor in such a system; thus most industry observers were not at all surprised at the demise of AHSC's distributor competitors. Today there are no other national distributors left in AHSC's market.

Most observers are less aware of what happened to the suppliers (like our chemical manufacturer) who tried to bypass

AHSC and sell direct to the hospitals. By the late 1970s or early 1980s, hospitals expected to be able to order suppliers by computer. Our supplier offered its own terminals, but they were not satisfactory; what the customers really wanted was to have the supplier's products handled by ASAP. Our supplier had no choice; dealing with AHSC became the only option. Since then it has had to lay off part of its direct sales force. Furthermore, AHSC is a much tougher negotiator than the hospital customers used to be, and as a result the manufacturer's margins have plummeted under the new arrangement.

What was the full impact of ASAP on this supplier? This supplier's relationships with its customers have been transformed under its nose. Its managers were shocked to discover that a distributor had so much control over a supplier and, what was worse, that the distributor could pull off such a coup without the supplier's even knowing what was happening until it was too late.

Consider the forces at work in this example. Hospital customers were facing pressures to control costs; AHSC-supplied technology not only helped reduce AHSC's administrative costs but also helped control its customers' inventory and ordering costs. Similarly, in an increasingly competitive market, AHSC, its distributor competitors, and the manufacturers were all searching for competitive edges; AHSC was smart enough to find one based on technology. Old patterns in channels of distribution began to change, in part because technology enabled the change to take place.

The ASAP story is particularly instructive because many organizations today face the same marketplace forces, only in even stronger forms. In the remainder of this chapter, we will first examine major forces affecting or threatening almost all companies today. We look at them as general business changes, although they all are at least partly brought about by information and information technology and in fact they help create the Information Imperative. We will then look at what information and information technology can do to help provide solutions.

Challenges for Today's Organization

Here is a list of the major external and internal forces that affect or challenge organizations today:

1. Customer relationships are changing.

2. Channels of distribution are changing.

3. Competitors are changing.

4. The organizations themselves are changing.

5. Technology is changing.

These changes are central to the Information Imperative. Because of them, astute managers are asking disturbing questions:

> Could some of these changes be leaving us out of touch with our customers, even though we've always assumed we were especially close to those customers?

> Could changes in our market be making us vulnerable to the actions of competitors, suppliers, distributors, or other channel partners? Could one of them impose change on us because they understand what is happening better than we do?

For an increasing number of managers in a variety of companies, the frightening answer to at least some of these questions is yes. Changing customer relationships, power shifts in channels of distribution, and changes in the basis of competition are threatening numerous firms. Transformation is hard enough to deal with when we see it coming; if we are unprepared, it can be fatal to the business.

Throughout the rest of this chapter, we will use the term *marketing task* broadly, to mean all that a company does to identify and satisfy the needs of its customers. Figure 3–1 shows the internal and external pressures affecting the marketing task in both industrial and consumer companies. Changes in cus-

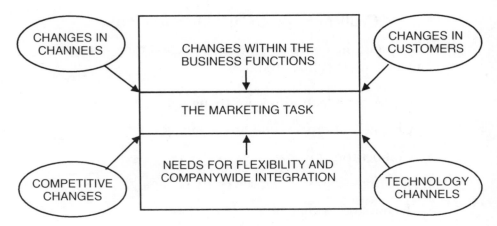

Figure 3–1. *Pressures Affecting the Marketing Task*

tomers, channels of distribution, competitors, and technology are external sources of pressure; internal pressures include changes within the traditional functions (such as production and marketing), as well as companywide requirements for flexibility and integrated strategies and actions. While we discuss both internal and external pressures throughout this chapter, we will emphasize external pressures. (Chapter 4 examines changes in individual functions.)

1. Changing Customer Relationships

Improved information and information technology help empower industrial customers to demand more. Industrial customers are demanding shorter and more reliable delivery schedules, especially if they are moving to just-in-time systems for their own production and inventory management. They are also demanding higher quality and products built or at least assembled to meet their individual needs. They are intolerant of long waits for maintenance and other service. At the same time they are demanding more, however, they are also pressuring suppliers to keep prices low.

Information and information technology have empowered

industrial customers and have thereby become the enablers of change. They allow customers to measure and monitor their suppliers' performance far more closely than was possible in the past. Thus, customers can document when they are receiving good service and when they are not, when a supplier's performance improves and when it slips. At the same time, information and information technology are providing these customers with benefits of efficiency, effectiveness, and transformation for themselves. As they improve their own businesses, the customers expect the cooperation and contribution of their suppliers.

The U.S. automakers provide a good example. As the three major automakers strive to become more successful in the face of strong foreign competition, they are transforming the nature of their relationships with their suppliers. After decades in which it emphasized the importance of having alternate suppliers (in order to play one against another), Detroit has changed direction. The auto manufacturers are now saying that in the future they will use fewer suppliers—from whom they will expect more. Information and information technology, both within each of the three major automakers and between each firm and its suppliers, are central in enabling the change.

Industrial customers display diverse behavior; they demand marketing approaches that fit their particular needs. Some customers make individual purchase decisions on the basis of immediate inducements to buy—today's price, speedy availability, specific product features, and the like. An example might be the traditional purchaser of carbon steel who today could select the supplier with the lowest price or best delivery and be very happy with the results, and yet tomorrow could select a different supplier on the basis of tomorrow's price or tomorrow's delivery. Such customers are transaction buyers; they are best served by transaction marketing strategies.

Other customers are relationship buyers; they want strong, close ties with sellers and expect vendors to act accordingly. Most customers purchasing large, mainframe computers provide a good example. It is terribly difficult for these customers to switch from one vendor to another after they have made an initial commitment; as a result, most such customers plan not to change. Instead, they rely on their chosen vendors to serve them both today and tomorrow. These customers buy not only on the basis of the vendor's immediate inducements; they often describe themselves as "buying the company" rather than buying the product. In serving relationship buyers, the seller must develop relationship marketing strategies.

We often see transaction and relationship buyers within the same marketplace, even for the same product. Particular customers may act as transaction buyers for some purchases, as relationship buyers for other purchases, and as something in between these two extremes for still other purchases.

Not surprisingly, relationship and transaction marketing strategies are fundamentally different from one another. Either can be executed successfully and profitably where appropriate, but it is often disastrous to attempt one strategy when the other strategy fits better. Sellers must understand the types of buyers they have (or could choose, or could create) before they choose strategies for them. Information and information technology help companies obtain the needed understanding. They also can help companies create stronger ties with their customers. For example, ASAP created stronger ties between AHSC and its customers by transforming transaction customers into relationship customers.[1]

Consumer markets are increasingly segmented. Years ago, it was relatively simple to identify consumers as either "department-store shoppers" or "discount-store shoppers." No longer. Today, most consumers shop in most kinds of stores. Consequently, stores can no longer rely on the patronage of loyal customers. Instead, they must provide individually attractive shopping and buying opportunities to individual customers.

A typical cart in a supermarket illustrates the change. One cart will likely contain everything from no-name generic products to gourmet luxuries. Consumers shop differently for different products and different occasions. One consumer may buy generic peanut butter and specially baked bread; another may buy custom-ground peanut butter and ordinary bread. Further, the same consumer may purchase differently even in the same product category, depending on the intended user or the planned usage occasion.

It seems obvious that many organizations can no longer segment their customers based on yesterday's simple demographic variables. Today's markets require meticulous market segmentation—careful segmentation based on a deeper understanding of how and why the consumer buys—for product development, for targeted advertising campaigns and consumer promotions by manufacturers, and for store-level merchandising decisions by retailers. Such refined segmentation would be impossible without improved information and advanced information technology.

2. Changes in Channels of Distribution

> Major power shifts are occurring in channels of distribution.

The effects of increasing diversity of customer behavior (both industrial and consumer) are exacerbated by changes in distribution channels. For many firms, power shifts like that of the American Hospital Supply story, while intriguing, are also deeply frightening. Distributors and other intermediaries are gaining substantial power in some industries and are losing power or even disappearing in other industries. It is often difficult to tell before the fact which pattern will prevail in a particular situation. The ASAP example illustrates one pattern. By contrast, manufacturers in other industries are using information technology to assume traditional distributor functions and to increase their direct business with customers; in these industries distributors fear that they will be weakened or eliminated. There are cities in the United States, for example, where hardware distributors have been eliminated (and the channels supplying retail hardware stores have thus been simplified).

As we saw in chapter 1, retailers' use of scanner technology is changing distribution channels in fundamental ways. This technology allows retailers to capture volumes of purchase data at low or no marginal cost. More than twenty years after the development of the concept, it has finally become feasible to monitor direct product profitability—the economic contribution of an item after consideration of its selling price, initial cost, handling cost, space requirements, and other factors. Scanners are helping retailers become increasingly knowledgeable, sophisticated—and powerful. As a result, packaged goods manufacturers are finding it increasingly difficult to convince retailers to accept products that do not perform well or to try promotions that do not pay. These manufacturers are having to adjust to a world in which retailers—and even brokers—know more about the performance of their products than do their own managers and sales representatives.

3. Hungrier Competitors

Competition appears increasingly fierce in many industries.

Making matters worse, many companies are facing fiercer and fiercer competition. Most firms today are stressing the need to be more customer- or marketing-driven. While the idea is little more than a slogan for some, others are working hard to understand their customers and markets better and are then using their understanding to compete more effectively (for example, through better customer selection or through the design of targeted product offerings).

Similarly, companies are striving to understand the expanded business value of information and information technology and to use that understanding to gain competitive advantage. American Hospital Supply had the luxury of being able to make its strategic move gradually, largely because its competitors were not technically ready to copy the early ASAP systems. Most of today's companies do not have that luxury. They should be convinced that at least some of their competitors (and perhaps some of their channel partners as well) are working diligently to find their own "ASAP systems"—their own transforming strategic moves based on technology. They are working to uncover ways to change the bases of competition within their own industries or to increase their power relative to that of their customers, suppliers, or rivals.

Competitors today appear increasingly willing and able to execute tactical competitive moves, such as changing their products, delivery, or service (or any of a number of other variables), to win patronage from customers. Information and information technology are key, both in providing the understanding with which companies can identify promising moves and in helping companies manage the enormous changes resulting from those moves.

4. Internal Changes

Skyrocketing costs of individual business functions represent one of the key internal pressures. For now let us consider marketing and sales as an example. (The following chapter considers a wider range of functions.) To many business managers, the costs of a direct sales force appear headed skyward. At the same time, salespeople are required to master more and more information. Media costs are staggering. Most consumer marketers consider trade promotion costs out of control.

Cost is one of the factors forcing businesses to improve their understanding of their customers and potential customers, their product benefits, and their competition in order to run their individual functions efficiently and effectively. It also motivates them to monitor and control their uses of all their resources. In other words, as chapter 4 elaborates, cost is one force pressuring managers to obtain the benefits of efficiency, effectiveness, and transformation for individual business functions or departments.

Targeting, flexibility, and integration are required if companies are to deal with the key external and internal forces of change. Increasing customer diversity, changing channels of distribution, competition, and costs require that today's company target its actions to particular situations. Homogeneous, across-the-board strategies and tactics are not likely to succeed with customers who demand products and services that fit their specific needs—especially when competitors are so often willing and able to provide the requested tailoring.

Further, the continuing pace of change requires that today's companies be flexible—that they be able to adapt and then adapt again. It is not enough to find the solution once (even when it is a transformational solution that benefits the whole organization); it is essential to find ways to adapt and update the solution over time.

Finally, success in today's marketplace increasingly demands that companies integrate all of their activities: sales, manufacturing, marketing analysis, service, and others. Customers will not settle for a good product with poor service. Or they won't settle for a good product with excellent technical service if uncertainty in the seller's production schedule makes shipment dates unpredictable.

The Benefit/Beneficiary Matrix is a tool for identifying not only applications and procedures that improve individual or department-level performance, but also applications that improve the organization's efficiency and effectiveness, and, in some cases, that transform the organization as a whole. Demanding customers, new and/or changing channel partners, and hungry competitors require that companies employ sound, integrated strategies in response to customer demands; piecemeal solutions executed department by department will not suffice.

5. *Changes in Technology*

As we have noted before, information and information technology are key forces creating the major challenges we have discussed. They are also important components of most solutions. Information and information technology allow substantial improvements in the performance of the marketing task. When incorporated into marketplace offerings, they often add significant value: they can strengthen links between sellers and buyers; they can even transform transaction buyers into relationship buyers. If these potential contributions of technology are the good news, then the bad news is that customers, competitors, and channel partners all know about technology's promise. Few companies have the luxury of deciding whether or not to make changes and improvements; they must do so in order to survive. That news, good and bad, is the essence of the Information Imperative.

Looking to Information and Information Technology for Solutions

To respond successfully to these profound challenges, managers must first understand the challenges, then design actions and processes, and finally implement solutions. Information and information technology, as we have already suggested, are central to all three steps. Figure 3–2 summarizes their key contributions.

> Businesses need information for understanding first, before they can design and implement successful competitive strategies.

Successful businesses respond to today's challenges with sound marketing—identifying customers' needs in depth and satisfying those needs profitably. Before they can proceed to designing and implementing effective solutions, managers need to understand

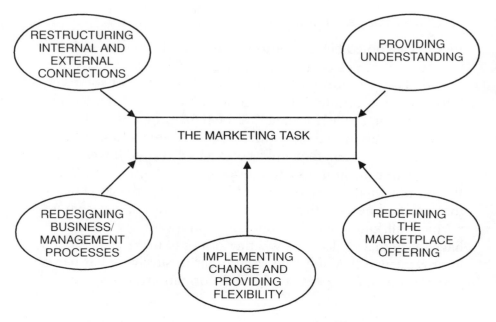

Figure 3–2. *Solutions Based on Information and Information Technology*

the forces of change; carefully selected, presented, and maintained information provides the required understanding.

What form should that information take? Huge, static databases are not the answer; they are expensive, take too long to construct, and almost invariably fail to provide the right information at the right time. Instead, managers need flexible, adaptable, action-oriented databases.

In order to illustrate the power that improved understanding can bring, let us examine a consumer packaged goods company that has an important new product in test market in four cities. To date, sales in all four cities have been somewhat disappointing, and the executives of the firm are trying to decide what corrective actions to take.

Naturally, the appropriate actions depend on the specifics of consumers' responses. For this discussion, let us consider two extreme possibilities:

> *Case I:* Many consumers in the test cities have tried the new product, but very few have repurchased it or plan a repeat purchase.
>
> *Case II:* Few consumers have tried the product, but a healthy majority of those who tried it like it and plan to repurchase the product.

In Case I, it appears that the product is poor; in Case II, the product seems fine, but there appear to be problems with the communication program for its introduction.

As the distinction between Case I and Case II suggests, differentiating trial purchases from repeat purchases is critical for a packaged goods business. Consequently, companies in this industry have long used market research techniques to trace the purchases of individual consumers over time, to identify trial and repeat purchases. In the past they relied mostly on consumer diary panels for this information. Typically the packaged goods company (or another business) would form a panel by asking a sample of consumers (usually housewives) to write down—after every shopping trip—details about what they bought, what prices they paid, and so forth. In return for this information, the firm would give the participants small gifts from time to time.

Although they constituted the best tool available at the time, these manual diaries posed serious problems. An obvious one concerns the accuracy of the data; participants had to record purchases after every trip—and had to do so accurately—in order to provide good data. A second, less obvious problem was more serious and grew increasingly so as more and more women began working outside the home. Marketers realized that people willing to write down all of their purchases were not likely to be representative consumers. So, even if the data were accurate, they might be substantially misleading.

Today the Behaviorscan service from Information Resources Inc. [IRI] has used technology to solve the old problems with panels. IRI selected test cities and then ensured that there were scanners in all the relevant stores in those cities (giving the equipment away if appropriate). The company then signed up a panel of consumers who were willing to present encoded cards at the checkout register each time they shopped. The scanners automatically captured the data about what individual consumers bought, who they were, what they paid, and so on.

This Behaviorscan service is obviously much more accurate than its manual ancestor. True, consumers can forget to present their cards—but using the card is probably relatively easy for most (and the participants are still given gifts as incentives for using the cards). Even more important, the range of consumers willing to participate (and therefore register their preferences voluntarily) is far wider than the range of consumers willing to keep the manual diaries.

In short, packaged goods companies are able to obtain more accurate data from a more representative sample of consumers— certainly a good example of how information and information technology can be the vehicle for improved understanding. Information and information technology are or can be similarly important for understanding in virtually all industries.

Places to Look for Solutions

The American Hospital Supply ASAP system, which we discussed earlier in this chapter, falls into several of what we will call "solution categories." (These categories are not meant to be mutually exclusive). In the remainder of this chapter, we will discuss these three solution categories, which are based on information and information technology:

1. Redefining the marketplace offering

2. Restructuring internal and external connections

3. Redesigning business/management processes

Redefining the Marketplace Offering. Increasingly, the designing and implementation steps involve changing the marketplace offering. Many of today's businesses are using information and information technology to change their products and services. For example, it is possible to build intelligence into some products. Thus, a chip may incorporate diagnostic ability into equipment. Often, information and information technology allow solutions tailored to the needs of specific customers; they can, for example, provide the flexibility to assemble custom products from standard building blocks, or they can fit delivery schedules to particular customers. Businesses can also use information and information technology to improve quality and other aspects of performance while also controlling costs.

American Hospital Supply's ASAP system falls into this category. Before ASAP, AHSC distributed a wide assortment of hospital products to its customers. After ASAP, AHSC still distributed a wide assortment of products, but it also offered services that helped customers control inventory and ordering costs. It had changed its offering from one of distribution to one of increasing the hospital's productivity, with distribution as a component.

Rand McNally, a company that has traditionally made maps, has also used information and information technology to redefine its marketplace offering. Over the years, Rand McNally amassed a wealth of data for its mapmaking business. In recent years, Rand McNally has automated much of its mapmaking, thereby obtaining benefits of efficiency and effectiveness. More important, however, the company is now able to combine general databases with customer-specific data to produce new products, a wide and flexibile set of customized map offerings. In essence, Rand McNally used information technology to transform itself from a seller of relatively standardized map products to a customized map producer (and maintainer of relevant data for its customers). In that process, it appears to have strengthened its links with its customers.

Restructuring Internal and External Connections. Other keys to successful designing and implementing are the connections among a company and its customers and channel partners and the connections within the company. Information and information technology can form the basis for closer links.

ASAP obviously fits in this category as well. It illustrates computerized transaction systems that strengthen links between sellers and buyers. As another example, toll-free 800 numbers strengthen the link between buyers and sellers by providing product and service assistance to buyers of products ranging from PC software to industrial equipment to consumer appliances.

Here is another example. ACME Paint, a (fictitious) auto body paint supplier that has traditionally relied on steady business from the major auto manufacturers in Detroit, is facing two seemingly incompatible demands. The big three's emphasis on efficiency, effectiveness, and transformation is putting increasing pressure on ACME to serve them well (or not at all), yet today's competitive situation makes it imperative not to incur unnecessary costs.

ACME's managers want to provide the automakers with the pigments they need when they need them. But demand is uncertain because nobody can accurately predict what colors the consumers will choose. It is impractical for ACME to increase inventories and still control working capital costs.

How can information and information technology help ACME managers? To gain a competitive advantage, they could try to obtain the earliest available real information about what colors are being ordered. If they can convince the automakers to let them plug into the car-order information systems, they can get the information they need to adapt their production schedules and thus serve their customers better without having to stockpile inventory. In this case, links between the customer and the supplier promise to strengthen the relationship between the two.

Internal links (such as department-to-department links) can move a business past inadequate, piecemeal business actions and, instead, provide integrated responses that squarely address marketing and other business problems. As an example, suppose Omega Corporation is a (fictitious) company that sells a technically complicated product in an industrial market. Omega's managers have begun to realize that their customers are upset by the lack of coordination among sales representatives, technical experts, and others from Omega. Customers are hinting that Omega had better begin responding in a coordinated way—and do so quickly—or they will take their business elsewhere.

How can Omega use information and information technology to bring the organization's full knowledge and resources to their customers when needed? The General Electric Plastic Division found one way. Its customers often needed help in selecting plastics suited to particular applications. The selection required a large base of technical knowledge; in the past, it had frequently taken sales people weeks to obtain the necessary information and return to the customer with recommendations. GE put its technical database on-line and made it accessible to the field sales representatives. As a result, an individual salesperson could recommend options that reflected the depth of GE's technical knowledge—and could do so quickly.

Redesigning Business/Management Processes. The responses we have been describing require carefully designed processes to capture information efficiently, to get it to the right managers at the right time, and to coordinate the efforts of the different parts of the business. Once again, information and information technology are central to the audit and redesign of business processes. Improved information can guide business decisions, obviously. As chapter 2 explains, prototyping can provide learn-and adapt-as-you-go implementation for addressing critical but unstructured business problems. Decision support systems are information systems designed for particular managers and specifically designed to prompt actions by those managers; they can provide managers with tailored, flexible, action-oriented tools.

Imagine that the sales vice president for a large industrial company has just received what could be some very good news. A competitor is having problems with a key product under some conditions in customers' operations. The manager's firm's product does not seem to have the problem. How can this company take advantage of the situation?

The industrial firm will have more options if it already has an effective customer information system in place. To underscore this point, let us consider two cases:

Case I: The (unfortunately typical) company does not maintain good account profiles.

Case II: The company has computerized account profiles of all important customers and prospects, including sales histories (if any), information on purchases (including competitive ones), and so on.

In Case I, our vice president tries to take advantage of the situation by sending a memo to the field offices, but this response is hardly as effective as one would like. This method assumes that individual salespeople will *read* the memo, determine whether it applies to their customers, and act accordingly. It also presumes that the sales representatives will *know* which prospects might be having problems with the competitive product and should be called immediately. It is improbable that all the salespeople would have such information at their fingertips or in their heads! In all likelihood, some salespeople will be able to take at least partial advantage of the situation; others will not.

Contrast Case I with Case II. With automated account profiles, a simple search can produce a targeted list for each salesperson and each district, identifying current customers and prospects who should receive immediate sales calls. Case II constitutes a big difference from Case I and, in all probability, a profitable one.

The final key to success in today's challenging business environment is to accept—even welcome—change as a way of life. Understanding, designing, and implementing must be continuing, responsive processes. Information and information technology can help provide ongoing education and exploration, continuing audit and design, and flexible, integrated responses.

Picking Priorities: Deciding Where to Start

Most companies cannot implement all the feasible responses to today's challenges. Therefore they must pick their shots, or select their priorities, carefully. They must understand the needs of varied customers, competitive pressures, and other forces well enough to identify the most promising actions, and then they must execute their critical moves well. The exhibit "How Information Technology Provides Specific Responses to Today's Challenges" lists common, specific approaches for meeting today's challenges. The following chapter elaborates on some of the responses for individual business functions.

The most appropriate immediate approach will vary from company to company. One company may decide that changes in customers' needs, in competition, and in sales costs make its central (and urgent) challenge the efficient and effective use of its field sales force; that company would most likely conduct an audit and redesign of the tasks of sales representatives and sales

How Information Technology
Provides Specific Responses
to Today's Challenges

1. Customer information systems: usable, flexible, adaptable

2. Customer relationship analysis and ongoing audit

3. Meticulous market segmentation and active tracking

4. Channel assessment: analysis, possible redesign, and on-going audit

5. Competitive information systems: usable, flexible, adaptable

6. Product/service redefinition through information and technology

7. Education about information and information technology

8. Individual business function: audit and redesign (for example, audit and redesign of the sales task)

9. Management process: audit and redesign (for example, for production management or for top executives)

10. Decision support systems for managers: tailored, action-oriented, flexible

managers. By contrast, another company might be most concerned immediately with power changes in channels of distribution; it would therefore benefit from a channel assessment, redesign, and audit. In these cases, as in hundreds of companies today, the key remains the same:

Understand, design, and implement. Face today's challenges squarely, using information and information technology to select and deploy successful actions for efficiency, effectiveness, and, especially, to transform the organization as a whole.

Note

1. For more complete discussions of transaction marketing and relationship marketing, see: Barbara Bund Jackson, *Winning and Keeping Industrial Customers: The Dynamics of Customer Relationships* (Lexington, Mass.: Lexington Books, 1985); and Barbara Bund Jackson, "Build Customer Relationships That Last," *Harvard Business Review* (November–December 1985): 120–28.

4
The Imperative for Each
Business Function

The Information Imperative affects the business function or department as well as the organization as a whole. As they work to achieve the goals of their companies, the individual functions will also use Domain III applications (see figure 2–10). They will apply the key tools of information and information technology for efficiency moves, for actions aimed at effectiveness, and, most important, for transformation.

In this chapter, we examine the Information Imperative for the function. We do not attempt an exhaustive examination of any particular function. Instead, we first summarize the forces that create the Information Imperative for business functions in general. We then consider the functions' classes of responses, all based in part on information and information technology. We illustrate those response categories with examples from a variety of specific functions.

Challenges for the Function

Challenges for the function are created primarily by the basic challenges to the organization. The business functions frequently feel special pressures to control costs and to coordinate more effectively with other departments.

Since the missions of the functions are to carry out the strategy and achieve the objectives of the organization as a whole, the challenges for today's departments are primarily the basic business challenges that we discussed in the preceding chapter. Changes in customers, in channels of distribution, in competitors, and in technology all create pressure for the business departments within a company.

Feeling pressures affecting the marketing task, general and functional managers today display heightened concern with the competitiveness of their companies. They worry about their understanding of the marketplace, about their market shares, and often, especially, about their own costs.

To some extent, this emphasis on cost reflects past slackness. Many managers believe that their organizations have been lax in the past, paying too little attention to efficiency, productivity, and control. To the extent that they are correct, those managers are behaving appropriately in emphasizing cost control.

To at least some extent, however, today's emphasis on cost may be less soundly based. A variety of observers believe that business is too focused on the short term and on easily quantifiable results. Such focus tends to emphasize control against a short-term budget. And because the business functions are generally controlled by such budgets and because they are usually called on to provide desired productivity improvements, they typically feel the brunt of programs for cost control.

Lest the reader conclude that we are advocating a lack of cost control we should elaborate. Cost control is of course important—especially in the face of hungry competitors and of increasing worldwide competition. Our major objection is to the use of too short a time horizon and too limited a perspective. In terms of the Benefit/Beneficiary Matrix, we worry that considerations of cost may lead managers to overemphasize efficiency moves and therefore to miss the ultimately more rewarding potential moves for effectiveness and transformation (which can turn out to be very efficient too). Excessive focus on cost may keep these managers in Domain I (see figure 2–8) and preclude their moving to Domain III. And excessive emphasis on cost can motivate managers to use inappropriate, narrow measures to evaluate investments in information technology or other business decisions. Chapter 2 argued that outmoded justification measures, such as the "headcount and keystroke" measures of Domain I, can preclude movement into the more advanced domains.

As an example, consider a manufacturing firm that sells a relatively broad line of products to other companies. Motivated by increasing foreign competition, the company's top management has announced a program of strict cost control. An initial study showed that the logistics function of the company was a large, although partly hidden, cost. The manager of that function was told to "cut all of the fat" from her department.

Sensibly, our manager's first moves were to review the overall corporate cost study and then to collect additional details about her own department. That analysis highlighted the company's eight warehouses as a promising target. First, the warehouses were expensive, consuming a significant share of the logistics budget. Second, several managers sensed that the warehouses were not adequately efficient. And finally, the logistics manager knew that other companies in a variety of industries were using information technology to improve their warehouse operations.

Accordingly, the logistics manager assigned two analysts to study the warehouses in detail and to identify efficiency moves. The project team also included one of the company's internal consultants and a member of its information systems group. The logistics manager herself participated actively, providing support and ideas.

Within two months the project team devised an impressive program for the warehouses. They completely redesigned the paper flow. The new more automated design could be implemented quickly with a combination of existing systems, a purchased software package, and a small amount of new programming.

The change would reduce the clerical staff, eliminate a few other warehouse jobs, and reduce required inventory levels. The team estimated that the warehouse costs would be reduced by 15–20 percent.

The program was implemented smoothly. Cost savings were in fact closer to 20 percent than to 15 percent. The logistics manager and the project team were lauded for their contribution to the corporate cost-cutting program.

In fact, however, the company has missed the boat. A different study with a broader scope would have revealed several important factors. The company's customers have improved their production processes substantially and, if motivated to do so, could provide our manufacturer with sound advance estimates of their requirements. The customers are also working to improve their purchasing, as part of their own programs for cost control. As a result, many would be highly interested in providing a manufacturer with advance warning of their requirements and with more regular orders in exchange for modest price concessions. And the manufacturer's own production processes have improved to the point that the manufacturer could serve such customers directly, *without any warehouses.*

To be sure, eliminating the warehouses would require initial investments in new information technology and procedures. After the adoption period, however, the streamlined distribution system would be both more efficient and more effective. (It might also be part of a wider transformation, changing the nature of the relationship between the manufacturer and its customers.) The project team missed the boat because it looked at too limited a picture and used too short a time horizon. Excessive emphasis on immediate efficiency led the members of the team to miss the opportunities for effectiveness and transformation—and longer-run efficiency, as well.

This example also highlights the increased need for different functions to coordinate with one another. Knowledge of advances in production (by the manufacturer and by its customers) might have prompted the team to identify the possibility of eliminating the warehouses. And designing and implementing such a program would require input and cooperation from marketing (about pricing) and from production (about its capabilities and about what information it would require from customers to make the new business system work).

Many of today's creative business solutions cut across traditional departmental boundaries. Even what is primarily a solution within one department, such as production, is often far more powerful and effective if it includes strong explicit links into other departments, such as sales and accounting. Information technology is frequently important to those links, and by the same token coordination among departments is often important to realizing the potential of information and information technology.

Categories of Solutions

The preceding chapter argued that information and information technology are key components of today's solutions to the overall challenges to the firm; they are also key to solutions for the specific function. We further argued that the major solution steps are first to understand the challenges, then to design actions and processes, and finally to implement solutions. These same three generic steps are also appropriate for the specific function, although of course the steps take somewhat different forms for the business as a whole and for the various departments. The remainder of this chapter examines each of the steps in turn, illustrating its application to the business function.

Understanding Today's Challenges

The first basic requirement for solutions for the functions is understanding—in the form of information about those functions together with conceptual models of how they do and could work.

Even within the particular business function we rarely today have the in-depth understanding that is required to meet the Information Imperative. Businesses and businesspeople often operate by habit, according to rules of thumb, or with methods that have been devised through practice and experimentation over extended periods. There is nothing wrong with using those habits, rules of thumb, and established methods, *provided that the assumptions underlying them and the conditions for which they were designed do not change substantially.*

The problem, of course, is that assumptions and conditions do change—in fact, they appear to be changing ever more rapidly. Business managers may realize (and complain) that the old rules of thumb and methods no longer work. What those managers generally lack are the frameworks or conceptual "models" to understand what the old methods assumed and why they worked in the past; such understanding would provide the foundation either for altering the methods to meet changed assumptions or for concluding that the methods are truly outmoded and should be abandoned.

Production is a good example of a function in which our current inadequate understanding is proving to be an increasing problem. In the United States today, production understanding is generally product understanding, not process understanding. Managers may have considerable data and sound understanding about the technology of specific products, about the individual production steps used to make those products, and about how the products behave in use. Rarely, however, do they have models or pictures of the overall production process: its components, their interrelations, potential problems and bottlenecks, the assumptions underlying the process.

Today's drive toward more automated production (toward

computer integrated manufacturing, or CIM, and its various relations) is highlighting this lack of process models. Not surprisingly, the business press has begun regular coverage of companies that are encountering serious problems in their automation efforts and are therefore scaling back or slowing down those efforts. Managers try to automate pieces of their production processes. But they do so without an adequate understanding of how the parts fit together—without understanding, for example, the problems that one step can create for later ones or the role of a later step in compensating for the lack of uniformity in an earlier one. Managers may also lack the understanding of the individual steps that is needed to manage and adjust those steps successfully in new and unfamiliar conditions. It is in fact not really surprising that automation efforts undertaken without such process understanding are not successful.

Pricing is frequently an area that demonstrates the importance of being explicit about assumptions. Businesses rarely keep good written records of the logic on which they base decisions, such as pricing decisions. Even more rarely do managers record the critical assumptions on which a decision, such as a pricing one, was based and which should therefore be monitored because *changes in the assumptions should trigger a reevaluation of the decision.*

Retrospective reconstruction of the logic and assumptions behind a decision generally simply does not work. For one thing, our recollection of why a decision was made is invariably affected by hindsight and by what we have learned since the time of the decision. In addition, there is often an even more obvious problem in reconstructing business logic. Managers are often rotated through a variety of assignments as their careers progress. When a pricing or other decision is reviewed, even a year or two after it was made, the managers responsible for the analysis and decision have frequently moved on to other jobs. Without written records of the logic and assumptions, the new managers have little chance of gaining a thorough understanding—and therefore a greatly reduced chance of reaching an optimal new decision.

The lessons of these pricing and production examples carry over to other business areas. A sound conceptual model of how a business function works is generally needed to identify changes in that function, especially the more creative moves aimed at effectiveness and transformation. And recording the detailed logic and assumptions behind business decisions is an important device to allow monitoring, adaptation, or even abandonment of those decisions, as appropriate. Sound understanding and record keeping are critical to Domain III management.

Avoiding the Pitfalls of Technology Push

> Lack of sound understanding is a common contributor to the problem of technology push. Good business models help avoid that problem.

Adequate understanding and sound conceptual models also help avoid excessive technology push (that is, driving from the technology rather than from the business need). Good knowledge of technology is of course important to meeting the Information Imperative—for suggesting creative options, for recognizing what is and what is not possible, and for implementing successfully. Unfortunately, however, excessive emphasis on the technology is a common problem for managers trying to realize the competitive power of information and information technology. Sound, successful technology moves, especially those in Domain III, must be driven from business needs and business understanding. Technology can both inspire and empower—but it should generally *not* lead.

The sales force frequently provides an illustration. We will consider Midwest Lubricants, Inc., a fictitious company created as an amalgam of several actual examples, both industrial marketers and consumer marketers. (Technology push is unfortunately common in the sales force at present.) Midwest sells diverse lubricants to a wide range of industrial users. The largest customers are served directly by Midwest's sales force. That sales force also calls on distributors of industrial supplies, who serve the remainder of Midwest's users. The sales force is compensated with a salary, a small commission, and a bonus.

Midwest's sales function faces pressures that affect many sales forces today. One problem is sheer cost. Each year the sales vice president glumly reads the published estimates of the cost of an average industrial sales call, and each year he concludes that his own company's average cost per call appears headed through the roof. A second problem is that, like many salespeople (both industrial and consumer), Midwest's representatives are being asked to master larger and larger amounts of information—about products and their uses and about the needs and histories of individual customer accounts. Some salespeople master the information considerably better than others. Finally, these expensive and increasingly knowledgeable salespeople appear to spend far too little time in actual customer contact. Too much of their time is consumed by filling in call reports, playing telephone tag with the home office to determine order status or for similar purposes, and doing other administrative work. Both the salespeople and their managers complain about the waste of time. The salespeople also complain that it is not nearly easy enough for them to obtain the home office's version of their individual sales volumes—both to see how they're doing and to check that their commissions are being calculated properly.

While Midwest's sales managers have been acutely aware of these problems, the impetus for using technology in the field sales force came from elsewhere. Sales representatives began to report that the salespeople of their largest competitor were carrying hand-held computers. No one was quite sure what the competitive representatives were actually doing with the handhelds, but, as is surprisingly common, the mere appearance of that technology spurred several Midwest salespeople and sales managers to demand a response. A champion emerged, and a task force was established to automate the field sales force.

An outside consultant recommended strongly that Midwest first build a picture or conceptual model of the desired strategic role of the sales function, that it then assess the fit of the current sales force and sales process with that desired role, and only then that it consider what, if anything, field automation could contribute. The task force demurred. The technology champion was pushing hard and was in a hurry. Hardware vendors were urging Midwest to proceed and were offering assistance in implementation. A study of the sales role would take time and would likely require an expenditure for consulting assistance. Instead, Midwest began distributing hand-helds to the field. The first application was to be the automation of call reporting, the most common source of complaints. The second would be a program to track and calculate commissions.

Eighteen months later the field automation project is in trouble. Training the salespeople proved more difficult than anticipated. Most of the representatives are now finally able to use the programs for call reporting and for commissions. They very much like the commission program. The problem is that the automation program has been expensive (in both time and money) but it does not appear to be producing much in the way of business results. Midwest managers have never been really clear or explicit about the purpose of call reporting in general, and automated call reporting is doing nothing more than saving a bit of time for the more proficient users. The commission program does make the sales force somewhat happier, but the total benefits of the project to date are hardly commensurate with its costs. And no one is quite sure what to do next with the new technology.

Better understanding at the outset could have produced a considerably happier result. Midwest's sales managers could have tackled the basic sales problems. They could have determined which tasks actually require the expensive sales force and which could be handled some other way (perhaps involving computer or communication technology). They could have identified ways to help salespeople to master information and, even more important, to use that information effectively. They could have worked for efficiency not only by speeding up existing sales tasks but also by redefining or eliminating some of those tasks. And they could have considered even broader strategic issues, such as the question of whether Midwest's distributors were worth their cost in margins or whether Midwest could use another distribution strategy instead. In short, Midwest's managers could have driven their efforts from business needs and sound business understanding rather than succumbing to technology push.

Places to Look for Solutions

The preceding chapter suggested three (not mutually exclusive) solution categories for the business as a whole: redefining the marketplace offering, restructuring internal and external connections, and redesigning business/management process. With some rewording to acknowledge the change of focus, these same three solution categories (still not mutually exclusive) are appropriate places in which to look for solutions for the business function.

Major solutions categories for particular business functions or departments are:

1. Redefining the department's output or results

2. Restructuring connections within the department or between the department and other functions in the business

3. Redesigning the processes for managing the function

Redefining Outputs. Increasingly, information and information technology allow the business functions to improve the outputs they provide to other functions and to customers and other out-siders. Those key tools help speed the response of the function. They often allow outputs to be tailored to particular conditions and needs. They may also help make a basic product or service easier to sell, easier to service, or more effective in use.

The human resource department provides one example. The days of standard rigid benefits packages are past in many firms. Instead, companies are moving toward flexible benefits. Many offer employees at least a few options: alternative medical cov-erages, for example. Some plan fully flexible programs, with em-ployees given "budgets" to spend on vacation, medical coverage, life insurance, and other benefits in combinations suited to their individual circumstances and preferences. Such complex-ity would surely not be manageable without information technology.

In the production area, information is increasingly enhancing the value of basic physical products to the sales force and other insiders and to the customer too. Production can provide sales with easy access to information about order status, allowing salespeople to respond quickly and efficiently to customer in-quiries. Other information can also be useful to customers. A chemical supplier has begun including in its shipments listings of the results of quality tests it performs; the information allows the supplier's customers to use the chemicals more effectively with fewer problems. Another manufacturer automatically no-tifies customers of planned and actual shipment dates. Note that both quality and shipment information are especially important to customers trying to adopt CIM.

Restructuring Internal and External Connections. Information and information technology help create smoother and more use-ful links among the parts of a specific function and between that function and others. Earlier in this chapter we discussed the problems inherent in trying to automate the pieces of a produc-tion operation without an adequate conceptual model of the process as a whole. Once they have such a model, managers can proceed to automate the appropriate parts of a possibly redefined process, tying the parts together and managing them as a whole with carefully selected information flows. The preceding section suggested that production could smooth its connections with the sales function and with the customer by providing easily accessible order status information. As a further improvement, production could establish regular early information flows from salespeople and from customers to forecast future requirements and facilitate production scheduling and procurement.

The finance function illustrates a history of several changes in connections and in departmental results that have been made possible by information technology. Typically the earliest adopter of computers in a company, finance (including accounting) usually has the longest such history.

Before the adoption of computers, finance departments were generally consumed with the basics of financial accounting—with closing the books and reporting to stockholders, with payroll, accounts payable, and accounts receivable. The first computer applications in finance simply made those basic functions more efficient and more accurate.

Later information technology moves in finance continued to provide additional efficiency. For example, consider a company with area offices throughout the United States. The business's first information systems were centralized. Payroll, accounts payable, accounts receivable, and general ledger systems all ran at the headquarters data center. All vouchers, time cards, and other relevant pieces of paper were sent to headquarters for processing. More recently, the company has distributed the accounting systems to the area offices. Paper is processed in those offices; only needed summary information goes to headquarters. The basic business method has not changed, but the accounting flow is faster and smoother.

More important, the power of information technology combined with savings of time and effort from earlier efficiency moves allowed finance to become more effective and even to change its basic role. Finance managers were no longer consumed with financial accounting. Instead, they were able to devote resources to management accounting too—to helping managers in other departments understand the financial implications of past or potential business decisions and, on occasion, even to identifying promising business actions. In addition to reporting the past, finance became active in analysis for the future.

More recently, additional changes, some of them strongly related to technology, have been creating continued substantial changes in the role of finance. In part because of increased business education, managers in other departments have better and better financial skills themselves. Personal computers, spreadsheet programs, fourth-generation languages, and other tools of end-user computing allow such managers to perform their own analyses. And those managers want to consider a wider range of information, not just financial data. Increasingly, they are making measurement an integral part of their management processes—not the separate contribution of a different department.

For companies in which these trends are especially strong, the finance department will shrink. Finance will retain its role in financial (or public) accounting and in determining standards and procedures related to such accounting. Finance will not, however, continue as the primary provider of management accounting. It is likely to be only the provider of the financial data and infrastructure needed by others for management accounting. And finance may also assume a role as the collector and consolidator of nonfinancial data for analysis.

Notice how central information technology has been to this series of changes in the role of finance and in the relations between finance and other functions. (Chapter 7 will argue that one possible future model for the information systems function is very similar to the role just sketched for finance. Information systems may become the keeper of the technical standards and infrastructure, but ownership and management of applications and even of hardware may shift to the user departments.)

Redesigning Management Processes. As was true for the business as a whole, today's responses for the business functions require carefully designed processes to capture information efficiently, to get it to the right managers at the right times, and to help turn its implications into effective actions. Again, information and information technology are central to the audit and redesign of management processes.

The following chapter explores in some detail the Information Imperative as it applies to the roles and processes of the company's top executives. This section instead considers examples of management processes within the functions.

First we consider the (repair) service department of an equipment manufacturer. Service departments have been something of a challenge for such companies, whose managers have debated the proper role for service: cost center or profit center, separate business or adjunct to the main equipment business, and so on. Many managers have become increasingly aware of the importance of service in winning and keeping the patronage of customers as a potential source of profit (or at least as an opportunity to control cost), and in providing useful information to sales, product development, and other areas.

Early in the information era (in the 1970s) Kodak used information technology in a new process for managing service for all of its equipment businesses. Kodak assigned each piece of equipment a unique identifying number. Customers requiring service called a local service center and gave Kodak the equipment number. The dispatcher then immediately accessed the entire computerized service history for the equipment. That history and the caller's description of the current symptoms allowed the dispatcher to decide what action to take. If a service person were dispatched, he or she called the local service center after examining the equipment. The center then advised the service person what parts were needed, whether they were on his or her truck (whose parts inventory was known by the computer), and if not, whether the needed parts were on some nearby truck. Each night, the trucks returned to the service center, where the computer directed the replenishing of their parts inventories. And the center relayed summary information to headquarters for use by product development and other interested functions.

As another example, we consider the marketing function of a consumer packaged goods company. The preceding chapters described some of the challenges facing such companies. Their consumers are increasingly segmented. Their retailers are using information to alter the power balance. Retailing decisions are increasingly made locally rather than for an entire chain. Trade promotion often appears out of control. Media costs are staggering.

For many such marketers, part of a response will involve more targeted marketing management and decisions. There will be more regional advertising. Trade promotions will be monitored and managed customer by customer. Merchandising programs and sales arguments will be tailored more to local conditions—sometimes to the individual store.

Information technology will be key to managing the explosion of data implied by such changes. And before that technology can be truly effective, careful audit and some redesign of the marketing management processes will be needed to help managers to focus on key issues and decisions and to manage rather than drown in the detail.

Implementing Successful Solutions

Chapter 2 stressed the importance of using appropriate implementation strategy and tactics in Domain III moves with information and information technology. Careful, thorough implementation is important in actions focused on the business function just as it is in other moves.

One important step is not to underestimate the complexity of the task.

This and the earlier chapters emphasized the value of driving technology-based solutions for the functions from business needs and sound understanding. We have argued the value of searching for broader, more fundamental opportunities as well as for simpler, more straightforward ones. Such moves are hard, to say the least. Successful CIM is proving extremely difficult. So is implementing a major shift in the role of a sales force. Such moves may well be worth the effort and if so should often be adopted, but it is a serious error to underestimate their difficulty.

Finally, adaptation and monitoring remain crucial for successful implementation. Adaptive implementation (often involving prototyping) is frequently needed, especially for more important and fundamental changes. It is also critical to determine *(early)* the measures that will indicate how and whether a move is working and will allow adaptation of the initial implementation plan. And it is critical to use those measures actively in guiding implementation and thus to realize the potential of information-based solutions.

5
Technology on the Executive Desk

T he Information Imperative also represents a personal imperative: the threat and opportunity of information technology and systems in the work and career of the individual. This chapter considers the individual executive. We have already stressed that many executives are unschooled in the potential of Domain III (see figure 2–10) applications to provide competitive and transforming benefit; the two preceding chapters provide a first step in educating them about applications, especially Domain III applications for organizations and departments. In fact, however, many executives remain uncomfortable with (and perhaps unsuccessful at) even Domain II (see figure 2–9) applications and the personal use of computers. In this chapter, we analyze the phenomenon of executive personal computing, with particular emphasis on the overlooked and subtle issues and barriers that determine whether and when executives will be responsive to information technology. We explore the benefit that technology-enabled changes in executive work can bring.

The Phenomenon of Executive Personal Computing

The past several years have seen substantial discussion of senior executives' possible use of computers:

The *Wall Street Journal* ran a special section entitled "Technology in the Workplace," partly devoted to executive computing.[1]

An article entitled "The CEO Goes On-line" appeared in the *Harvard Business Review.*[2]

Businessweek ran a special report on the subject, and *Fortune* magazine ran a cover photograph (and story) about Ben Heinemann, then chairman of Northwest Industries, shown seated at the controls of his executive information system.

Advertisements for "lap-top" portable computers portray fast-track senior executives using their products aboard executive jets, and ads for "executive decision support" software packages are now commonplace in business publications aimed at general management.

Players ranging from Big-8 accounting firms to individual entrepreneurs have packaged hands-on computer literacy courses for senior executives.

Nevertheless, the vast majority of senior managers have had no direct personal experience with computing. They readily admit lack of knowledge, and they also admit having questions:

Is a computer *really* for me?

What would I do with it?

Real managers don't do spreadsheets, do they?

Wouldn't using a computer make me appear to be looking over the shoulders of my management team?

If I started to use a computer, what would the reaction be within my organization?

The experiences of some executives who *have* had direct involvement with personal computing cannot be considered successes. Some candidly confess that the personal computers on their credenzas are expensive (and often dusty) paperweights. Others may point with pride to their terminal and then proceed to give eloquent testimony to its practical value by forgetting how to start it or by getting stuck somewhere in the menu.

Finally, for a select number of executives the personal computer has become an absolutely essential tool. These executives maintain that technology has worked nothing short of a major transformation in the pace, effectiveness, and competitiveness of their management activity and in the work of their management teams.

These findings, by themselves, are not particularly surprising. At one time in the past, senior executives would have expressed the same kinds of views, in roughly the same proportions, about the telephone. More challenging is the task of determining the *reasons* that some executives experience little or no value from the computer while others see it as an extraordinarily powerful force in reshaping management. This chapter identifies and explains those reasons and provides tools for assessing the potential value (or absence of it) of executive information systems (or EISs, as they're called). We next present three disguised case histories as the basis for that discussion.

Three Case Histories

A False Start at Mammoth Oil Corporation

Mammoth Oil is a large, profitable, integrated energy company with a stable, experienced executive team and a reputation for being well run. Mammoth's chief executive was convinced that an executive information system would prove invaluable to him and to Mammoth's senior officers. He believed that Mammoth Oil had entered an era far more dynamic and threatening than its executive team recognized. He reasoned that an EIS would expose the team to different kinds of information and would provoke them to think more critically, analytically, and imaginatively. Further, electronic mail, word processing, calendaring, and other office support functions would boost management productivity.

The initial version of an executive information system was developed with remarkable speed. Personal computers were installed in the offices of the senior executives. The EIS made available a broad variety of internal and external reports and databases and also functions such as calendaring and electronic messaging. Each of the executives received careful orientation and training.

System usage skyrocketed the first week after installation. It then went down just as quickly. The EIS remained brain-dead until the end of the fiscal quarter, when the executives turned on their tubes to read the financial results and then shut them off again. Only the CEO was making significant use of the system. There were several attempts to revive usage. Then, all of the office computers except the CEO's were removed, to the chagrin and frustration of the CEO and to the relief of his senior management team.

Adoption at Progressive Paper Corporation

Progressive Paper Corporation was a profitable, well-managed company; it strove to be first to adopt state-of-the-art techniques, technologies, and policies in every aspect of its business. So aggressive was Progressive Paper in adopting new ideas that some critics regarded its management as "gimmick-happy."

It came as no surprise to Progressive's top managers when one day a personal computer showed up in the office of their president and CEO. Nor were they surprised when several months later, the president and vice president of Information Systems announced plans for a network of personal computers, both office computers and portable/home computers, to support the top management team. There were a few initial grumbles about fear of machines, inability to type, and disruptions of office furniture arrangements, but they faded quickly in the face of the CEO's persistence and charm. Before long, many of the executives proudly showed visitors their executive information system and even demonstrated some proficiency in using it.

Management at Progressive changed very little, however. Informal working communications among the executives continued to take place largely in the hallways; the form and content of meetings, decision processes, and other operating mechanisms continued as usual. Several years later, an efficiency-oriented "corporate fiscal fitness" program included a survey of Progressive's top management. They were asked, "We are spending x million dollars per year on the Executive Information System; how would you feel if this system were discontinued as part of our overall cost reduction activity?" The response from the executive team was overwhelmingly neutral.

Transformation at Fortune Enterprises

Fortune Enterprises was a leading company in a very large industry. For most of its existence, Fortune had enjoyed unchallenged market leadership and spectacular growth in sales and earnings. Eventually, the company's success attracted niche competition, and revenues and profits began to level off. Galvanized into action by serious and mounting competitive threats, Fortune undertook a massive management reorganization and strategy redirection. It shifted from a functional and geographic organization to one consisting of much smaller strategic business units (SBUs). Each SBU was headed by an individual general manager, who was given control of most elements of the SBU's worldwide profit equation.

The new SBUs faced an extraordinary set of pressures:

Most SBU managers had traditional functional backgrounds and had never been general managers

Each SBU was asked to satisfy the needs of a particular market and customer set rather than to focus on a specific product

Performance would be judged on external market measures rather than on traditional internal ones

The new general managers were uncertain of their grasp of their newly defined businesses in an environment of great change and competitive challenge. They also discovered to their dismay that the individual work-styles to which they had become accustomed over many years were downright dysfunctional in the organization. In particular, Fortune's "let's have a meeting" style of communication was clearly unworkable. The general managers and their teams now had to coordinate activities with substantial numbers of shared research and manufacturing organizations. They also had to communicate with one another from various locations around the world (and in a wide range of time zones).

At the same time as the massive reorganization, Fortune began a project to develop executive support systems for the business units' management teams. The resultant systems provided the SBU teams with access to information necessary to shape and monitor their individual business strategies. These systems also provided a significant array of communication capabilities, including electronic mail and document transmission.

The business unit executives quickly realized that the executive information systems made the reorganization workable. Those systems made possible a dramatic change in the manner in which individual managers and teams of managers performed their jobs. In essence, the managers expected (at most) Domain II efficiency and effectiveness from the systems. They got efficiency, effectiveness, and also transformation of their work. Here are some of their comments:

Whenever we had a plant loading/sourcing meeting, 80 percent of the time was taken up by information exchange, leaving too little time for decision making. Then we'd have to reconvene to take action and the process would start over again. Our executive information system is changing all that. *Before* the meeting we can now view the key information we'll need. The system is increasingly used *in* the meeting itself to portray alternatives, and there's now plenty of time for deciding on actions.

We've discovered that many fewer physical meetings need to be held. Important discussions and decisions can be accomplished over the phone or through electronic messaging, now that we have concurrent access to the same information and tools for analysis.

Our executive information system has dramatically affected our "managerial velocity." Things happen faster, communication is crisper. Sure it's taken some getting used to, but we're one or two years ahead in our strategic action programs compared with where we'd be managing the old way.

Our secretarial and staff support needs are altogether different now. We're able to do far more with fewer, shared clerical staff. We can see significant impacts upon clerical salaries as well as office space requirements.

Conditions for Executive Success with I/T

Why did Fortune Enterprises realize both Domain II benefits and Domain III transformation for its managers, while Progressive Paper achieved at most only simple Domain II value and Mammoth Oil got nowhere? We believe that two basic factors differentiating these cases also explain why executive information systems are or are not successful.

The impact of information systems on executives generally has almost nothing to do with the underlying information technology. Instead it is a function of two varieties of stress:

1. Stress on the business
2. Stress on individual work-styles

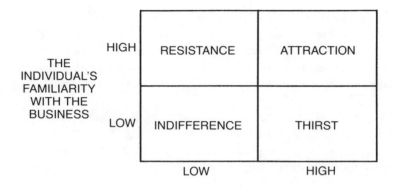

**CHANGE IN THE BUSINESS
AS PERCEIVED BY THE INDIVIDUAL**

Figure 5–1. *Key Factors Affecting Executives' Demand for Improved
Information Support*

Stress on the Business and Its Managers

As a reliable general rule, executive information systems produce little or no significant impact in the absence of stress on the business that is felt and acknowledged as serious by its managers.

Figure 5–1 shows two dimensions of business stress from the perspective of the individual executive. The first is the executive's perception of the degree of change occurring in the company and its industry. The second is the degree to which the individual executive is familiar with the business and comfortable about his or her own understanding or "feel" for it. (Such comfort is usually built during considerable experience.) Figure 5–1 shows how the effects of the two forces combine. High perceived change and low familiarity, for example, combine to create "thirst," or great receptivity to the benefits of information systems.

A closer examination of our three cases will illustrate the

point further. At Mammoth Oil, all the members of the executive team were very high on the "familiarity" dimension of figure 5–1. All had been in the petroleum business for at least twenty years, and most had spent a long time at Mammoth. The position of various executives along the "perceived change" dimension, however, varied considerably. The CEO sponsor of the system believed that the petroleum industry was undergoing enormous change, requiring a new way of looking at the business and an aggressive pursuit of additional insight and understanding. He wanted the executive information system to help him and the managers to think more analytically and critically about what was really going on in the industry. The nub of the problem with regard to the EIS was that very few of the other senior executives (and *none* of the opinion leaders among that group) shared the CEO's perception of the degree of industry change and required changes in management learning and behavior. In the terminology of figure 5–1, the CEO was in the region of "attraction" bordering on "thirst," while the remainder of his senior team was in the "resistance" cell. At Progressive Paper Corporation, everyone on the management team (including the CEO) was "high" in familiarity with the business and "low" in perceived change. The resistance to the EIS predicted for this case by figure 5–1 was in fact overridden by a cultural norm that valued trying new things and being perceived as "state of the art" and by the CEO's persuasion and advocacy. Although that norm overcame resistance, it could not alone create the environment for a successful EIS. By contrast, at Fortune Enterprises the disorientation and stress associated with the major strategic moves thrust essentially all of the SBU executives into a condition of "thirst" for improved information.

Stress on Work-styles

Stress on individual and organizational work-styles also generates demand or need for support for executives. In the Fortune Enterprise example, stress on the executives came from redefi-

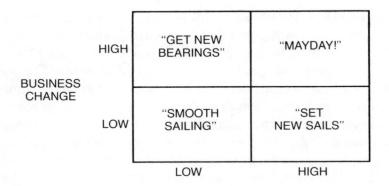

WORK-STYLE STRESS

Figure 5–2. *Work-style Breakdowns Amplify Need for Executive Information Systems*

nition of the business, from new performance criteria, and from the fact that old working habits suddenly became impractical. In other words, the individuals faced new business missions in their jobs, and at the same time they found their old work-styles totally inappropriate. Figure 5–2 depicts the effects of the two types of stress. The executives at Fortune Enterprises were in the territory labeled "Mayday." Had the reorganization not produced a breakdown of their traditional work-styles, they would instead have been merely trying to "get new bearings," seeking new information to understand and manage in their redefined missions and businesses. Or, had the reorganization left the business definitions intact but played havoc with old work-styles (as major geographic reorganizations of sales regions and spans of control can do), these individuals would need to "set new sails" to adjust to the new conditions. As it happened, *both* sets of changes were imposed on Fortune Enterprises' executives simultaneously.

Compare the Fortune Enterprises case with that of Progressive Paper. Recall that Progressive's executives were under no significant stress for either business or work-style change ("smooth sailing"); it's no wonder, therefore, that nothing much happened when they got executive workstations. Without stress there is no perception of need. EIS at Progressive was technology in search of a problem.

Executives in different regions of figure 5–2 need different types of support systems. Moving from "Low" to "High" along the business-change dimension increases the manager's need for facts or data to help build understanding and insight—a new mental model of the business. Moving along the work-style–breakdown dimension from "Low" to "High" generally increases the manager's need for support in communication and in specific business tasks. Put another way, movement toward greater business change produces an appetite for information retrieval, data manipulation, or simulation capabilities typical of what are called *decision support systems*. Movement toward greater work-style stress creates a need for the types of support typically found in *office systems* technology: text processing, electronic mail, and document sharing.

Successful Implementation of EIS

"This stuff isn't for kids!" recently remarked a manager responsible for the implementation of an executive information system effort that had successfully launched an initial set of applications.

His point was simply this: to create an executive information system that *truly makes a difference* in the activities and processes of management requires very wise, experienced, savvy, and skilled management along a surprising number of fronts—practically none of which are technical. This reality is in sharp contrast to the prevailing, naive view of many general managers and information systems managers. They believe that the central EIS implementation tasks are (1) defining the contents of the "executive database" (typically delegated to the controller's and planning staffs), (2) selecting the "friendliest" EIS software package (delegated to the I/S professionals), and (3) training "users."

Entire books are now being written on the implementation of executive information systems, so we will touch on just a small set of essentials. Our experience points to three necessary conditions:

1. An influential senior executive sponsor with a clear stake, desire, purpose, and vision for the EIS

2. A mechanism to achieve initial focus, clarity, and commitment around the types of information needed to make the system meet its intended purpose

3. A vehicle for EIS form, content, and usage to *evolve* apace with the *organizational* capacity to provide the needed information and adjust or adapt to the impact of those new flows and uses of the information by executives

A Vision of Success

Our experience has shown us that transformational executive information systems are created within the context of "a vision of success" held by an influential and strongly committed senior executive. The Fortune Enterprises case provides an example. One of Fortune's top executives was firmly convinced about two things: first, that new information would be absolutely essential to the ability of Fortune's SBU executives to perform as the new strategy and organization demanded; second, that the Fortune organization was already seriously overloaded with personnel engaged in information handling at all levels. He could see that the organizational and strategy changes would create substantial pressure for *additional* information-handling staff but realized that if Fortune were to be cost-competitive no growth in staff could be permitted; in fact, reductions of between 25 and 30 percent in staff personnel would be required. He had a vision of success: executive information systems would eventually let managers access needed information directly, rather than through layers of staff. This vision of success required substantial learning and adjustment by executives and a complete redefinition of the role of staff. It was a powerful vision, however, and together with the top executive's leadership and commitment it drove the changes at Fortune.

Focusing Managers' Information Needs

The majority of organizations building executive systems employ "supply side" thinking. That is, they focus activities on gathering a very large mass of internal and external data about the business and assembling it into an information utility complete with a wide variety of data access, manipulation, display, and communications tools for use by executives. This approach is commonly taken because executives' requests for information are notoriously unpredictable. However, most systems created in this fashion are generally not regarded as particularly useful by the executives for whom they were built. They also tend to be large, cumbersome, and expensive.

A more successful result emerges when the problem is redefined as that of focusing managers' information demand rather than as that of finding different ways to mobilize information supply. The process of focusing information demand begins with achieving focus, clarity, and commitment on the part of the management team to those few business activities that are truly critical to the success of the business. The emphasis here is on business *actions:* managers define just what it is that they must do to achieve their business goals. The most critical information needs are then defined as those required to support the key business actions. The critical success factors process (outlined briefly in figure 5–3 and explained more fully in an article by Jack Rockart)[3] is a powerful and proven technique for achieving the needed focus, clarity, and commitment with regard to business

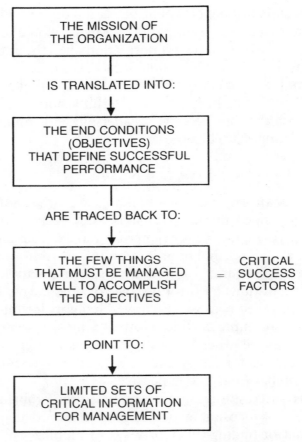

Figure 5–3. *The Critical Success Factors Process*

actions. With that process, an individual manager or a management team—even one that is being pulled in many directions by the imperatives of the business—can articulate and agree to a limited set of factors (five to eight) that will be central to the organization's success. Once defined, those critical success factors provide a beginning point for the evolution of executive information systems.

Evolutionary Development with Prototyping

Executive systems of the "paperweight" variety are *built and installed;* executive systems that transform are the product of *evolution.* Their development is not a process of construction by information systems people who are following executives' requests—or, more often, trying to guess what would be useful. Rather, their development is a process of learning by executives—individually at first and then collectively.

The starting point is the list of critical success factors (CSFs). The next step is to define broadly the types of information likely to support these critical actions and processes. For example, the critical success factors process and subsequent analysis might highlight such needs as improved market segmentation, analysis of market and customer profitability, closer measurement and control of the new product development process, and more thorough measurement of critical activities in new product rollouts.

While these areas of need may be rather clearly defined by the CSF process, however, the form and content of the needed information and the ways managers will actually use that information to support thinking and decision making remain undefined. We believe the definition *cannot* be complete at the outset. Evolutionary development is needed. Prototyping (introduced in chapter 2) is the appropriate tool.

Figure 5–4 outlines the prototyping concept. A manager learns what sort of information support is truly appropriate to his or her needs through direct trial-and-error experience in using a series of small-scale (but gradually more refined) prototype versions of the intended system. At the same time, the organization evolves its capability to provide access to the needed information. And this evolution of supply is the trickiest and most often mishandled part.

Everyone readily grasps the "try it—fix it—try it" nature of the side of prototyping that helps evolve the demand. Most overlook the fact that creating direct access to new types of data for new

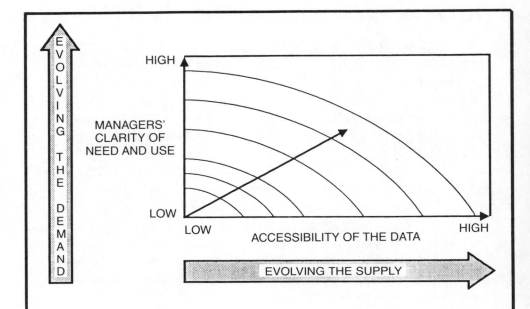

What has to evolve:

- Understanding of the true business need

- Understanding of what information is most meaningful—and how best to display it

- Management processes to provide a context for use of the information

- Mechanics—and organizational support for
  – New varieties of information
  – New patterns of information flow

Figure 5–4. *Prototyping*

types of users (senior managers) creates acute discomfort throughout an organization, and that it is as critical for the *organization* to evolve an accommodation with these changes as it is for the *executives* to evolve their own needs and uses for the information.

The central idea is that the development of executive information systems is not an exercise in specification and implementation; it is a process of learning on the part of the manager and of the organization.

Successful Executive Systems

Building successful information systems for executive use is a considerable challenge. Stress on the business and stress on the individual work-styles create the prerequisite conditions for success. A vision of success, clear focusing of managers' information needs, and evolutionary development with prototyping can then bring Domain III benefits to the roles of individual executives.

Notes

1. E. C. Gottschalk, "Executive Computing" in "Technology in the Workplace," *Wall Street Journal*, September 16, 1985, pp. 23cff.
2. J. F. Rockart and M. E. Treacy, "The CEO Goes On-Line," *Harvard Business Review* (January–February 1982):82–88. See also T. P. Gerrity and J. F. Rockart, "End User Computing: Are You a Leader or a Laggard?" *Sloan Management Review* (Summer 1986):25–34.
3. J. F. Rockart, "Chief Executives Define Their Own Data Needs," *Harvard Business Review* (March–April 1979):81–93.

6

Individual Transformation in the Workplace

Too often managers focus the information lens at the organizational and functional level and ignore the benefits of
information and technology for particular jobs or for the people
in those jobs. In fact, when we do think of individuals, it's often
as obstacles rather than opportunities. The result is mediocre
systems that meet no one's needs well.

In the past, a functional bias was inevitable. Information systems were expensive, technically complex, and difficult to build.
Tailoring them for individuals was out of the question. The rise
of end-user technologies has made focus on the individual possible and economically feasible.

The preceding chapter discussed Domain III benefits that focused on individual executives. Executive systems are important
and valuable, to be sure. The majority of workers are not executives, however. Information technology applications can bring
transformational benefits to the roles of middle- and lower-level
managers, professionals, and other individuals.

Focusing on the individual employee in considering information technology is consistent with other current trends. The
business world is currently emphasizing the importance of the
individual. Many business schools have begun courses on entrepreneurship—on harnessing the energy and creativity of an individual to start new businesses. Large companies stress internal
entrepreneurship, or "intrapreneurship." Training budgets are
rising. Individuals in manufacturing plants, working in small

teams, are being encouraged to rotate among several jobs. Small "tiger teams" are being given responsibility to develop new products. And so on.

In this chapter we will begin with examples of technology-enabled individual transformations. We will then describe major types of such transformations. Next, we will consider possible negative individual effects of information technology—and, therefore, the types of applications to avoid. Finally, we will describe factors that encourage the realization of Domain III benefits for individuals.

Kinds of Individual Transformations

Examples of Transformations

Information technology is transforming individual jobs:

- An insurance sales agent uses an expert system to provide financial planning advice to customers, for a fee. The agent is able to recommend particular financial products with much greater credibility. His sales show a dramatic rise in the first year with the system.

- An inventory clerk is given access to a terminal for a plant's computer integrated manufacturing (CIM) system. With the permission of his supervisor, he begins to investigate the causes of production delays and to correct problem situations. His title is eventually changed to "Production Expediter."

- An engineer in a chemical company discovers that a new computer-aided design (CAD) system makes her phenomenally more productive. Instead of spending most of her time drafting variations on older designs, she can now spend about 60 percent of her effort searching for significant improvements in the production process. She is able to find several sources of greater efficiency in the plant; she also finds her work much more interesting and decides to stop her search for a better job.

- When his company faces heightened competition based on consistent product quality, a manufacturing manager identifies new ways of monitoring quality for his company's main product. The resulting application measures "upstream" factors that have never been systematically measured before, including the composition of chemicals used in the manufacturing process. Analysis of the finished product shows much higher quality after the application is installed.

- In a major bank, an expert system for credit analysis is created for a senior loan officer, whose territory was a developing region abroad where the bank intends to make substantial loans. Cost-benefit analysis for the system indicates a savings in loan loss reserves of 5–10 percent for that region.

- In a large telecommunications firm, a system is developed to help a product manager select the best locations in which to place a particular type of credit-card telephone. The company has now deployed large numbers of phones to these chosen sites, where they are a significant source of revenue.

All of these examples show transformation. While in each case there were efficiency and/or effectiveness benefits in performing the job, the job itself has also changed. And although the individual's functional unit and the entire organization benefit from these changes, the primary change is in the individual job.

Categories of Transformations

Broader information technology applications aimed at functions or organizations also affect individuals. The individual benefits of such larger applications and of individual ones fall into four categories:

1. Those enabling the individual to assume new responsibilities or activities

2. Those enabling the individual to extend previous work

3. Those changing the individual him- or herself

4. Those providing benefits to the individual per se

Breakthrough Transformations. Information technology can enable individuals to undertake entirely new job responsibilities or activities, effectively creating new jobs. A creative breakthrough is frequently necessary to identify the new responsibilities. The breakthrough can come either from the individual whose job is to be changed or from someone else—a manager or systems analyst, for example.

Perhaps the most common of such transformations occurs when secretaries or administrative assistants assume the responsibility for creating and managing a key set of data. While the transformation of secretaries into database managers is not strategic in conventional terms, affected managers tell us that there are substantial benefits for the departments involved. There are also many examples of breakthrough changes at higher levels of the company involving professionals and managers.

We have found in general that breakthrough transformations occur in environments where individuals or their managers have access to information technology. Because secretaries often work with personal computers, for example, it is not surprising that they frequently create innovative applications on them. We have also found that the discovery of entirely new functions is more likely to occur when the people looking for transformation opportunities understand the old job environment well. It is difficult for a systems analyst, for example, to come into a department and propose entirely new ways for individuals to meet the business needs served by that department.

The major obstacle to breakthrough transformations is a rigid, bureaucratic organization that expects employees to continue doing the job that they were assigned. If there is no openness to individual growth and no process for recognizing and rewarding innovation and change that come from individuals, there will be few job breakthroughs.

Extensions of Previous Work. Some individual benefits are extensions of previous work. These benefits involve greater efficiency or effectiveness but are of such magnitude that they constitute a change in kind rather than degree.

Again, there are many examples. Roger Smith of General Motors envisioned a twelvefold increase in productivity on some engineering assignments from future computer applications.[1] (While recent reports about GM indicate that the implementation of such applications is proving more difficult than was first expected, such goals are feasible in some cases.) Scientists who use computers to simulate experiments, rather than actually conduct them, are able to realize enormous gains in productivity. Even the use of very conventional on-line transaction processing systems may lead to job transformation for clerks and administrative personnel. In such examples, there is more going on than the automation and speed-up of work activities. Some change in work process is also involved: the elimination of some steps in the process, or the use of different personnel, or a reduction in the resources required for the job.

These radical gains in efficiency and effectiveness can be even more controversial than changes involving entirely new responsibilities. Skeptics question their feasibility. Others suspect (sometimes with reason) that layoffs or cutbacks will follow the implementation of the information technology or that drastic increases in the pace of work will be required. If individuals are to embrace the potential of technology to bring transformation to their jobs, they must have a basic trust in the organization and in its interest in their welfare.

Changing the Individual. Some types of individual transformation involve not the job but the individual performing it. For example, we've been told the story of Joe, who is only a few years from retirement. Until recently he was just going through the motions. But since he got a computer, he's been coming in early and staying late. His manager has noticed a marked improvement in the quality of his work.

What are the reasons for this type of individual transformation? The view of Sherry Turkle is that when individuals work with computers, they begin to think of themselves as more powerful and capable, they understand work tasks better, and they think of new ways to do their work.[2] Further, technology can decrease or eliminate tedious, repetitive activities and enhance creative aspects of a job. The holder of that job *should* think more of him- or herself.

Such benefits of information technology are especially difficult to measure or prove. Yet the benefits to an organization of a competent, motivated, and creative work force are not in question. And, fortunately, this type of transformation seems to happen without much organizational intervention. All it requires is individual access to the technology, time, and the resources to experiment.

Benefits to the Individual Per Se. Information technology can benefit the individual per se, changing the relationship between the individual's home life and work life. While "telecommuting" is not suitable for all types of jobs or all work situations, it can sometimes work well. Some individuals are more productive when they work at home and find that they work longer hours overall. Working at home can limit time away from work because of illness, bearing and raising children, and the irritating details of life (such as staying at home for appliance repair calls). Home work can also bring entirely new groups into the labor force. For example, several insurance companies have used information technology to enable people at home to process claims. We expect that companies will continue to experiment with home work, learning its advantages and disadvantages in the process.

Negative Individual Impacts

> Information technology applications can also have substantial negative impacts on individuals. They can create dissatisfaction and can reduce or eliminate the characteristics of a job that employees value.

Critics worry about the potential of information technology to create negative transformations in individual jobs. Their concern has proved to be justified in many cases, and it shows no signs of abating. We do not intend to review the overall arguments of the critics. Instead, we describe here the types of applications most likely to lead to negative individual transformations. Knowing the types of dysfunctional applications makes it much easier to bring about employee satisfaction—by changing the application, communicating its purpose better, or abandoning a dysfunctional application before it consumes excessive resources. We consider four types:

Information Technology Applications Intended Primarily to Increase the Level of Monitoring and Control of Employees. In our experience, the benefits of such increased control are almost always offset by employee dissatisfaction and resulting poor performance and high turnover rates. Automated monitoring of the quantity of work leads to lowered emphasis on quality. As Shoshana Zuboff points out, increased monitoring leads to decreased initiative and creativity in employees.[3] Greater control may be acceptable in the context of a system that has positive transformational benefit overall. As an objective in itself, however, it is usually badly ill advised.

Applications That Remove or Drastically Reduce the Component of Thinking or Judgment in a Job. Most information technology applications thus far have, at best, supplemented human intelligence. In the few cases where the computer has replaced employee judgment, employees have usually rebelled. Perhaps even worse than open rebellion is the deadening effect of the absence of thinking in a job. With expert systems making the removal of judgment from individual jobs a real possibility, this type of negative impact should be a matter of concern for the future.

Applications That Remove All the Social Aspects of a Job. While there may be productivity benefits from telecommuting or from changing the physical location of a job, such moves require care. Much of our enjoyment of jobs and work is based on the opportunity to interact socially with fellow employees. Social contact need not be constant, but it cannot be totally absent.

Applications That Remove All Tangibility from a Job. Zuboff notes that we are accustomed in our jobs to dealing with tangible objects. When real objects are replaced by their representations on a computer screen, many people become confused and disoriented.[4] This problem can be eased to some degree by the proper design of user interfaces (such as those that replicate the appearance of a form) or by the creative use of graphics to depict tangible objects.

Fostering Individual Transformation

Domain II (see figure 2–9) and Domain III (see figure 2–10) benefits focused on individuals are potentially powerful and valuable. They are not easy to realize, however. Fostering individual transformation requires attention and management.

We suggest four major steps in fostering individual transformation:

1. Create a climate conducive to individual innovation.

2. Rethink information systems planning processes.

3. Identify and support individually oriented applications.

4. Target particular individuals for special information systems treatment.

Creating a Climate for Innovation

Simply put, an ideal environment for the exploration and use of information technology by individuals eliminates the negative and affirms the positive. Part of the positive involves the overall organizational climate; another part relates more specifically to I/S, the information systems function.

An overall need is an organizational culture that encourages and rewards individual growth and role expansion. If such a climate does not exist already, its creation is a task for the organization's most senior management. The human resources function is also involved in establishing evaluation and reward procedures that encourage individual transformation. Such cultural change is a long-term process with complex ramifications.[5]

An important part of a conducive overall culture is trust— trust that technology will be used to benefit individual employees rather than to harm them. Establishing such trust is a long-term process and is the responsibility of senior management. Trust can be fostered by always attempting to find new and better jobs for workers displaced by technology, by consulting with employees early about technology-induced job changes, and by sharing with employees any additional profits resulting from job transformation.

A positive climate for individual transformation also involves education. Such education should familiarize individuals with the concept of individual transformation, the general capabilities of information and technology in business, and the skills necessary to use basic tools.

Widespread, easy access to end-user technologies is another element of the positive environment. We find that people will generally not create innovative new computer applications if they must travel some distance—perhaps even down the hall—to get to the computer. Similarly, if the computer is available only at certain hours of the day, or if it must be shared with three or four other employees, the individual may not bother. Even when the individual is not creating the transforming application him- or herself, the new system may not be used if access to the technology is difficult. If an organization wants to encourage individual transformation, it must commit itself to making modern, powerful technology available in large quantities at the place where it will be applied—the individual's desk or workstation.

Rethinking Information Systems Planning Processes

We find that to date most ideas for Domain III information technology applications have bubbled up from individuals within the organization rather than coming from top-down planning processes. While, as the next chapter will elaborate, we believe that top-down planning should become better at finding opportunities for transformation, we also believe that it is important for the company to create an environment in which innovative ideas can bubble up. Management's technology vision and its view of technology's role in achieving business change should be communicated widely and regularly throughout the organization. Individuals should be encouraged to participate in I/S planning by thinking up systems opportunities—for their organizations, their departments, and themselves.

Identifying and Supporting Applications

Part of this step involves growing a library of ideas for breakthrough applications to provide individual transformation or massive increases in efficiency and effectiveness. It involves communicating the contents of that library to encourage further innovation. Another part involves recognizing and encouraging opinion leaders in the area of technology. Individuals make up their minds about information technology and its applications much the way they decide about how to vote or what kind of car to buy. They may read and study, but they almost always also talk to colleagues and friends whom they consider knowledgeable. These opinion leaders will arise in departments and work groups no matter what the organization does. A sophisticated organization can cultivate them, ensure that their opinions are accurate, and use them as champions for individual transformation.

Individual transformation is also fostered by devoting special attention to implementation. We recommend establishing a process for monitoring the effects of information technology on organizational and individual behavior. The process should also provide recommendations for steps that will ease implementation and increase the likelihood of individual transformation. User participation and review are important to such a process.

Targeting Particular Individuals

In the past, information technology applications have frequently been made available to all (or many) employees without discrimination. While such a practice is admirably democratic, we believe it is mistaken.

Generic systems generally do not have nearly as much impact as tailored ones. But customizing and implementing an application for an individual require considerable effort. Businesses cannot afford tailored applications for everyone.

Who should receive special attention? First, employees who are already receptive to information and technology should be considered. They should be knowledgeable or willing to learn and able to spend time experimenting.

Second, the employees should be important enough to the organization to warrant the effort. The most valuable and productive employees in a department should be considered first. And the search should begin in the departments most critical to the organization's success.

Importance of the Individual

Although the transformation of organizations and functional units has been the subject of much recent attention, focusing on the individual is extremely important too.[6] First of all, the changes that affect organizations and departments involve individuals. Information technology applications are designed, developed, implemented, used, and maintained by individuals. Creative and innovative schemes for the application of information technology have failed because the necessary behavior changes in individuals were not recognized or managed. Therefore, even when an application seems to be aimed at broad organizational or functional benefit, considering and managing the individual job changes associated with its implementation are critical.

In addition, Domain II and especially Domain III information technology applications aimed at the individual are increasingly able to provide personal and business benefits. Organizations that intend to meet the Information Imperative cannot afford to ignore them.

Notes

1. R. B. Smith, "The 21st Century Corporation" (Speech to the Detroit Economics Club, September 9, 1985).

2. S. Turkle, *The Second Self: Computers and the Human Spirit* (New York: Simon and Schuster, 1984).

3. S. Zuboff, "New Worlds of Computer-Mediated Work," *Harvard Business Review* (September–October 1982):142–52.

4. Ibid.

5. For treatment of such change in manufacturing, for example, see R. E. Walton and G. I. Susman, "People Policies for the New Machines," *Harvard Business Review* (March–April 1987):98–106.

6. For additional discussion of this idea, see T. H. Davenport, "Individuals: The Neglected Element in I/S Strategy." *Indications* 3:6 (Cambridge, Mass., November 1986).

7
Integrating Information
Technology into the Business

I n the first six chapters of this book, we examined the implications of the Information Imperative for industries, then for
the affected businesses, functional units, and individual users.
We have not yet discussed the implications for the information
systems (I/S) function, which is the delivery vehicle for systems
in all three domains of applications for all levels of users.

In this chapter we will take a fresh look at the I/S function
in light of the Information Imperative. We will review the history
of the I/S function, a history that has been similar across industries and businesses within industries. We will discuss the requirements for shaping a new, revitalized I/S function for the
future and suggest two scenarios for the I/S function. Most important, we will search for answers to the critical question of
how organizations can smoothly and seamlessly integrate the
I/S function into their businesses.

Like the other chapters, chapter 7 was written for two types
of readers: (1) senior executives who want to know how to organize, manage, and use the I/S function to the company's best
advantage, and (2) I/S executives who want to know how their
departments and their careers will be affected by the implications of the Information Imperative and what steps to take in
the near term to prepare for the inevitable changes.

The I/S Function, the I/S Unit, and the Integration of I/S

It is important to distinguish between what we call the I/S function and the I/S unit. The I/S function consists of the activities that support and deliver information technology applications, regardless of where these I/S activities are performed. The I/S function can be completely centralized, completely decentralized, or something in between. The I/S *function* is made up of four basic activities:

1. Planning of applications and of technical infrastructure

2. Development of applications

3. Operations or use of developed applications

4. Technical service and support of applications, of the technical delivery infrastructure, and of users

The I/S *unit* is an organizational entity assigned to perform some or all of the four activities of the I/S function. Thus, when the I/S function is centralized, the central I/S unit performs most of the activities of the I/S function. When the I/S function is decentralized, the activities are spread throughout the organization.

To meet the challenge and opportunity of the Information Imperative, a key question for senior management is how to organize units of I/S so that the I/S function is seamlessly integrated into the business.

Seamless integration refers to a desirable state in which the appropriate (and sometimes different) organizational stakeholders (such as users, I/S specialists, and management) have sufficient knowledge and control of the technology, application systems, operations, and management of the I/S function to apply information technology easily and effectively.[1]

Most of today's business environments abound with potential barriers to the effective use of information technology for business purposes. The fact that information systems have a reputation for being complex and difficult to build and use is a barrier to integration. Organizational barriers, created when separate I/S units compete for vital pieces of the I/S function, impede integration. Communications barriers thwart integration efforts: users do not always communicate their needs well, and I/S professionals do not always communicate their capabilities well. More barriers between I/S specialists and users develop because these separate stakeholders often do not share the same skills, goals, or values. All of these barriers, and others, make it very difficult for information technology to be integrated smoothly into the operations of the business.

> The removal of barriers to the effective use of information technology should be an overriding goal for the I/S function and for the business itself.

In most instances, the goal of seamless integration will not be achieved without fundamental changes in the organizational structure of the I/S unit, including clarification of ownership of the I/S function by different I/S units.

A state of seamless integration requires understanding among all stakeholders about what information and technology are capable of doing (and are doing), on the one hand, and what the business must do to achieve its purposes and goals on the other. It implies that senior managers understand not only the potential contributions of information technology applications but also their own appropriate roles in achieving that potential. It suggests that all managers assume an active role in exploiting the critical resource of information technology for the business and that I/S professionals play an active role in suggesting opportunities and implementing applications successfully. Finally, the state of seamless integration highlights the need for standard but flexible processes that aid in creating and delivering applications and are capable of change.

Note that the desirable state of seamless integration does not in itself suggest a specific pattern of ownership of the I/S function by different units. That is, the goal of integration does not offer prescriptions about which stakeholders should perform particular activities of the I/S function.

> The essential management tasks in achieving seamless integration are to decide on the mission of the I/S function in the business and to place ownership of I/S functional activities appropriately to achieve the mission.

Integration has always been a desirable goal, but today it is more important—and more achievable—than ever before. For one thing, we are breaking down the technological barriers to integration. Today's information technology has evolved to such attractive levels of price-performance, reliability, and ease of use that sophisticated yet simple-to-use, affordable application systems are available directly to users. More important, integration of the I/S function into the business is increasingly becoming necessary for success in business.

In many companies, some degree of integration will take place regardless of who owns the I/S function. Many stakeholders are demanding access and ownership. If the barriers are not removed soon, they will be torn down. In our opinion, however, a planned movement toward integration will yield a more stable and effective integrated environment. Hence the next question becomes, how can organizations achieve integration?

The I/S department in many organizations has the mission of being a "business within a business," charged with functioning efficiently, recovering its costs (or even making a profit), providing reliable service, and generally keeping its customers (the users) happy. In some companies, this approach is taken to the extreme and the I/S department functions as a completely separate business unit or profit center serving both internal and external customers.

But the "business within a business" mission is, in many respects, the opposite of a mission based on integration. Its orientation is not the effective use of information technology for the larger organization's business purposes; rather, it emphasizes the individual department. Despite the integration barriers it creates, the "business within a business" approach may have been appropriate in the past and, for some, as a stage of evolution. Today, however, with integration an appropriate mission, the "business within a business" approach no longer works. We must search for new missions.

The Historical Development of the I/S Function

Options for new I/S missions will become clearer if we examine the historical development of the I/S function. In the twenty or so years that information technology has been employed in business, the function has changed dramatically, spurred by new technologies and new applications. Each new development seemed to confuse our understanding of what we had previously learned. Yet the evolution of the I/S function has been remarkably similar from one company to the next. By reviewing its history, we can understand the evolution of the I/S function to its present form and identify legacies that will be important to exploit or to change as managers seek new missions and new distributions of functional activities among units. (We recommend that each firm trace the evolution of its own I/S function.) In this section we will review the evolution of the I/S function up to 1987, using the three domains of the Benefit/Beneficiary Matrix to trace the history.[2]

The Domain I Period: Transaction Systems and Management Information

The first computer applications for businesses used batch-processing technology to automate manual processes. Initially these applications were in accounting, then in engineering and manufacturing, and subsequently were used to obtain management information for operational control and decision making. The benefits of the early computer applications were functional efficiency and, to an increasing degree, functional effectiveness, as information from the transaction systems was used by functional managers for reporting and control.

At first, the computer did not change the business process at all. Applications required no change in work behavior by the user/operator; users merely did by computer tasks they had previously done manually. Batch technology was the delivery resource, enhanced in the 1970s with on-line systems.

During this historical period, applications began to shift from automating transactions to improving the operations of a business. In accounting, for example, the computer began providing more effective management-oriented reports. In engineering, the computer began to be used in model building; in administration, word-processing systems were not only faster and more accurate, but they also added some new capabilities. Thus the applications began to provide effectiveness benefits, although still primarily for functional beneficiaries.

In their early history, Domain I applications required only an elemental form of *planning,* mainly project-by-project applications plans. Long-range planning consisted of estimating people and equipment capacities for one to two years.

Responsibility for defining these early applications systems generally fell to users within the affected departments. Typically, someone in finance took responsibility for accounting systems, someone in engineering for design-calculation applications, someone in manufacturing for process-control applications, and someone from the administrative services unit or the typing pool for word-processing applications.[3] Planning was easy in one sense because it could be done separately for each area. But because each area looked only at its own needs, the I/S function was split into several units owned by different parts of the organization.

At first, managers could relatively easily compare the savings (typically in headcount or labor time) from projects with the investments required (in programming time and any incremental hardware investment). Such comparisons allowed them to screen and rank projects. As the systems increasingly brought nonmeasurable benefits, however, and as some organizations centralized heretofore separate I/S units, benefits became more difficult to measure and priorities more difficult to set. Steering committees were established, with representatives from the different user functions. By the mid-1970s, however, most steering committees had become politicized and had lost their effectiveness. The result in all but a few organizations was that no comprehensive planning process survived.

Corresponding to the emphasis on project-by-project planning, the *development* activity in this period evolved to a high state of precision. Applications were developed in four steps: specification, design, programming, and installation. Specification required dialogue between a systems analyst and an end-user; it typically resulted in written "functional specs" for the proposed system. For example, specifications might include the number and nature of the transactions to be processed (so many checks to be processed or items to be inventoried). They might also include rough estimates of the cost of development and estimates of anticipated savings. The design step produced another, more technical set of specifications that divided the system into programming blocks or modules. A system design defined data requirements, data flow, and processing. Next, the programming took place, typically in COBOL, FORTRAN, or BASIC. The programming ended with testing and debugging the system, in parts and then as a whole. Finally, after testing and further refinement, the system was installed and users were trained. The ability to develop systems and manage projects did not come easily to computer specialists. Much of the early Domain I period was marked by trial and error learning, particularly for programming large systems and for managing projects effectively.

Operations and *technical support* activities were carried out by the functional unit that planned and built the system. Where centralized I/S units evolved, a contractual relationship was formed between users who had no I/S unit and the central I/S unit, which performed the development, operation, and support activities for a fee.

Where centralized I/S units evolved, almost invariably they reported to finance. Toward the end of the period many businesses began to consolidate telecommunications (including voice communications and data communications) and office automation under the direction of the central I/S unit.

Several important *legacies* remain from this early period of Domain I applications. The major legacy is the centralist mentality. Once they had won consolidation battles, some I/S managers came to believe that theirs was the only way to provide computing. Although that attitude is rarer today, we still see pockets of this mentality in many companies.

Although the centralized I/S unit was typical of the Domain I period, in some cases impatient users bought or developed their own information systems and formed their own units. The period ended with an uneasy peace between central I/S and users. Users demanded more and better applications; I/S simply couldn't keep up with the demand. The tension was explosive. It was the prelude to the dramatic events of the Domain II period.

In many cases, parts of the I/S function developed as separate "islands," which often could not be integrated technically. Assigning responsibility to different user departments for different types of applications encouraged this trend, as did assigning the individual I/S professional to support one particular user group. The legacy produced by this insular development is a major stumbling block to integration today.

The Domain I period also produced the legacy of ROI as a measure of application value. As was discussed in chapter 2, managers could count the number of keystrokes saved or the reduction in headcount from new systems. They could document how much money they would save by increasing productivity, reducing or displacing staff, or avoiding future costs. Unfortunately, this headcount and keystroke approach is inappropriate for applications in later domains. It has been difficult to dissuade executives from using the headcount and keystroke approach to measuring benefits. As a result, these executives have often rejected systems such as those designed to improve business operations or customer service; the improvements could not be quantified adequately. Even among those proposed projects that were accepted, many effectiveness benefits were discovered by accident rather than through advance planning.

Less a legacy than preview of what was to come, the Domain I period also presented the first dilemma over priorities: Which applications are most important to the business, those that increase efficiency or those that improve effectiveness? Which are more important, opportunities presented by new types of applications or the continuing benefits of some of the older systems?

As the I/S function grew and became more complex, specialists were added. New management skills and more intensive coordination were required. I/S managers could no longer be only good technicians or equipment managers; they were also required to manage increasing numbers of people. Many data processing experts lacked (and some still lack) the skills for managing a collection of specialists. In this situation left over from Domain I, we see the roots of the fundamental management problem that many I/S departments are wrestling with today.

Finally, the Domain I period saw the first evidence that older systems and practices could become inadequate. We have learned from these earlier experiences to build in safeguards—both human and electronic—to ensure that systems run smoothly and reliably. We have also learned not to expect systems to last indefinitely.

The Domain II Period: End-User Computing

The period from 1975 to 1982 was the watershed in computing history; Michael Hammer has described it as the "revolt of the masses." Microprocessor technology began in the early 1980s to bring compact, easy-to-use computers to individual desk tops. The "end-user computing" phenomenon offered enthusiastic users an attractive alternative to dependence on a centralized—and sometimes resented—I/S unit.

In this period applications brought individuals both efficiency and effectiveness benefits. A sales representative could use word processing to set up "boilerplate" letters and proposals that could then be tailored to fit the customer's needs and printed as original documents. The corporate planning analyst could use a spreadsheet to do "what if" analyses.

Spreadsheet programs for personal computers (PCs) were responsible for selling users on personal computing. In addition, users discovered numerous other software packages for personal computers. Kinds of analyses previously available only on large mainframes were now available for personal computers. So were graphics packages that changed the way users perceived and presented information.

The entry of the personal computer meant the death of the earlier I/S *planning* approach. Disgruntled users simply bought their own personal computers without much prior planning. Many organizations were overrun with technology. The total number of available computing cycles and memory capacity (although rarely measured) grew exponentially.

For a while, users' infatuation with personal computers offered the I/S department a respite from user demands and from the need to change its planning for systems. But users began to demand I/S support for PC use. Soon, I/S tried to exert some control over personal computer purchase and use. Sorting out priorities remained a challenge; I/S didn't have the budget to build new applications *and* maintain or upgrade existing systems *and* support end-user computing. Steering committees were no longer an effective way to establish priorities, yet user demands kept pouring in. During this chaotic period the I/S planning activity was in disarray. It remains so for many businesses today.

In retrospect we can see the problems of this period more clearly, but at the time neither senior management, user management, nor I/S management really understood or could suggest an alternative planning process. No single planning approach seemed to be directly applicable to I/S planning; no single methodology seemed to work well in all situations.

The period of Domain II saw the introduction of a new and fundamentally different process of *applications development*—namely, end-user development. Microprocessor technology made it possible for the users to develop their own systems. Friendlier versions of traditional programming languages (such as BASIC and PASCAL) made possible the first wave of end-user development. The advent of the electronic spreadsheet and fourth-generation languages (which were designed to be easy to use) offered even more opportunities for end-user development. While the four steps of the development activity (specification, design, programming, and installation) might still technically be involved when a product manager or financial planner used a spreadsheet, the formality and sequence were gone. The user generally doesn't recognize those steps.

Applications in this period were not confined to end-user development. As other applications grew in size and complexity, so too did the development process. Further, managers increasingly recognized the profound organizational impacts of information technology. Complex "change management" procedures grew up to accommodate these changes, but they often proved inadequate to the task. The Domain II period also saw increasing use of prototyping for developing applications that could not be defined completely at the outset. (Prototyping has proved especially useful for important Domain III applications.)

Until the appearance of personal computers, *operations* remained largely the responsibility of the centralized I/S unit. As users began to develop and manage their own applications, some operations tasks fell naturally to them. Others remained with the I/S department. In short, the new technologies brought about the dispersion of some development tasks and of some operations activities as well.

The period altered forever the uneasy relationship between users and the centralized I/S department. The I/S department lost much of its control, while users became both more sophisticated and more demanding. Many I/S departments gradually came to terms with end-user involvement; they looked for ways to support users. Many formed information centers to provide such support.

As the nature of applications became more diffuse and more subject to obsolescence in the Domain II period, *technical service and support* became correspondingly more expensive and difficult to provide. Over time, I/S professionals have had to improve their abilities to translate user needs into effective applications. They have had to learn to exercise leadership without appearing to dominate or control users. Their evolving role has come to require highly developed negotiating skills. As organizations move toward Domain III applications, the proliferation of hardware, software, and telecommunications equipment among users and the growing number of I/S specialists within user departments have resulted in some diffusion of even the service and support function.

Perhaps the greatest *legacy* from the Domain II period is the continuing tension between users and I/S. Early in the period personal computer users were the few mavericks who were frustrated with I/S. Subsequently the many PC users have charged ahead, much to the delight of vendors and in spite of the reservations of I/S.

As with many good things, however, there is a dark side to the proliferation of information technology. Without traditional care in testing, user-developed systems too often break down, leaving users and their departments vulnerable at critical times. These systems often lack proper documentation; errors can cause disastrous consequences and finding errors is difficult. For instance, a Florida construction firm was awarded a contract because it bid $5 million less than the next highest bidder. Only after they began the project did its managers discover that an error in their spreadsheet program had caused an expensive discrepancy between the estimated cost and the true cost. (The firm lost in a subsequent lawsuit against the software developer.)

As most companies begin to make the transition to the era of Domain III applications, ownership of the I/S function is shared among users and a central I/S department. Never before have the boundaries been so blurred; never has the need for integration been so apparent.

The Domain III Period

In the mid-1980s, organizations are making the transition from emphasis on end-user computing to emphasis on Domain III applications. As of this writing, it is difficult to say with certainty exactly how the era of Domain III will evolve, but there is no question that some discontinuity will exist between the periods of Domain II and Domain III, as it did in earlier transitions from one period to the next.

Once again, technology is making the transition possible. Improved information technology capabilities are creating what Jack Rockart calls the "wired society." Technology allows the linking together of departments, divisions, and branches of businesses; it also allows linking with outsiders such as customers and suppliers.

The transformational applications of information technology include internal applications and external applications. Both involve change: the internal transformational applications involve changes in job tasks, functional objectives, or the organization's business processes; the external applications involve changes in the way an organization represents itself to and deals with the outside world. In not much more than twenty years, we have reached an era in which applications can significantly change the work of individuals and functional units or transform entire organizations or industries.

To date, there are relatively few systems that transform entire organizations. Most that do exist have resulted from chance discoveries or have evolved from systems developed in earlier eras. But many organizations are seriously looking for strategic opportunities. It is just a matter of time (and not very much time) before businesses employ Domain III applications for competitive advantage. Though many organizations are still wrestling with applications from previous domains, it is the Domain III applications that will be most important in enabling businesses to respond to the Information Imperative.

As chapter 2 noted, one of the biggest Domain III challenges will be achieving the desired behavioral changes when transformational systems are implemented. User involvement will be crucial from development through implementation. Without the ideas, support, and commitment of users whose behavior will be affected (whether inside or outside the organization), successful implementation is virtually impossible. We can already predict that cooperation of I/S professionals and users will emerge as a major issue in this era. In short, the issue of the integration of the I/S function into the business must be addressed in the period of Domain III.

As organizations emphasize Domain III, old approaches to *planning,* especially long-range or "architecture planning," will prove largely inadequate. In the majority of cases today, both architecture and applications planning are driven by cost control and the inevitable "squeaky wheels." There seem to be three major reasons for the sorry state of I/S planning: (1) technological advances have expanded the potential for applications; (2) senior management still lacks a thorough understanding of what information technology can do (and is doing) for their organizations; and (3) I/S departments often still lack adequately strong leadership and management.

> Domain III planning must integrate systems planning into the business's strategic planning process.

We suspect that organizations will use different types of planning, with different time spans and different justification methods, to address the diverse array of I/S needs. We expect to see a mixture of very careful strategic planning for the long term and ad hoc planning to respond to day-to-day pressures.

As organizations enter this period, *development* has already become a combination of traditional development, prototyping, and end-user development. The ownership of the development function is spread among users and I/S specialists and will, without question, remain so.

> During the period of Domain III, we must find new methods and procedures to encourage this diffusion of development. At the same time we need clarification of roles and, for some purposes, standardized approaches.

Currently and increasingly, *operations* too belong to both users and I/S units. We expect this split to continue, further blurring the boundaries between the central I/S function and the users. Some operations will operate data centers, invisible to the users, running massive software twenty-four hours a day; the operations supporting today's ATM networks are an example.

Others will be day-to-day user operations of applications on personal computers. Once again,

> the issue with operations is integration—how to permit the dispersal of operations and still maintain control.

Technical service and support, like planning, development, and operations, has become diffused among I/S technical specialists and users. Keeping the technology running will continue to be one of the main tasks for technical service and support. With the growth and diffusion of the I/S function throughout the organization, the potential for inefficient or poor technical service and support is very real. Effective integration will necessarily involve clarifying the responsibility for these technical skills.

The support activity will also see increased responsibility for technology scanning. Today, many personal computer users take part in the scanning process. One can almost always find an "expert" within a department who makes it his or her job to keep up with the latest personal computer products and advancements. These individual "hackers" frequently know more about their particular technical systems than do the official technical support experts in the central I/S department.

Organizational ownership is a key issue in the period of Domain III. The I/S function is no longer a "business within a business," as we described it early in the chapter; instead, it is made up of some combination of technical experts, usually working from a centralized I/S department, and users. The questions most organizations face today are:

> Who will own the four activities that make up the I/S function?

> What combination of people will be responsible for planning, applications development, operations, and technical service and support in the future?

Citicorp may provide one prototype of an I/S function early in the period of Domain III applications. That function is decentralized, with expertise dispersed among employees throughout the organization (including the chairman, John Reed, himself). There are also pockets of centralized activities—data centers, development and operations activities belonging to particular functions—plus information centers and end-user computing. To the outside, it looks like dozens of I/S units and different pockets of influence. Yet in our opinion, Citicorp is an organization positioned to address the critical issues of integration and Domain III applications.

It is impossible to predict one scenario that will apply to all firms. Instead, we believe that

> ownership of the I/S function will be defined differently in different businesses and industries in the era of Domain III applications.

If an organization has emphasized user-driven computing during the Domain II era, its I/S department will likely shrink in size and stature and its I/S function will continue to be dispersed. If, on the other hand, an organization has traditionally relied on a strong, centralized I/S department, then the I/S function will likely remain largely centralized. For example, it is impossible to imagine that American Airlines' reservations systems could run on a network of personal computers. Conversely, it is difficult to imagine that a regional sales office should have to rely on a central mainframe computer to access and analyze data about its accounts. Diffusion may be the source of strength for some organizations and of weakness for others.

A Vision of the Future

The ambiguity over what the I/S function will look like in the future causes frustration to senior executives and I/S executives alike. While it is clear that the I/S function is in transition, moving toward a new mission and structure, it is still unclear what the structure should be. (The mission should be integration.) To add to the confusion, some of the greatest resistance to change is coming from the I/S professionals themselves—an issue that must be addressed before change can take place.

> The major difference in the Domain III period is that for the first time the I/S function will look significantly different in different businesses and industries. There will be no one scenario.

In the extreme, the function in some companies will be so diffused that there may not be a central department, while other companies will retain or rebuild a strong central department; the majority of cases, however, will fall somewhere between these extremes.

Recent conclusions based on research and prediction by PRISM (an applied research offering of Index Group and Hammer and Company) tell us that we can expect to see a clarification of the I/S roles in future years.[4] Central I/S professionals will strengthen their roles as keepers of the standards and architects of the infrastructure, having responsibility for the security and integrity of data and basic technology. Users, on the other hand, will continue to do what they do best, defining the business needs.

Will users be required to become more technical and to get involved in such activities as architecture planning? Will I/S professionals be forced to improve their understanding of business needs? How will users work with I/S? How will I/S support users? We still do not know the answers to these questions.

In a recent survey of users and I/S professionals, both groups agreed that I/S will remain responsible for highly technical activities but that users will become responsible for matching applications to their business needs and for actively participating in applications development.

The required transition cannot happen overnight. Users do not yet have the technical knowledge to match systems with needs, nor do they have the experience to develop sophisticated applications. We expect that I/S professionals will continue to provide a strong advisory and support role, and that the transition will take place in different ways and at different rates from business to business. We are certain that the process will not be easy.

While we believe that there are many ways in which the four I/S activities (planning, applications development, operations, and technical services and support) can be structured in the era of Domain III applications, we have chosen two extreme scenarios as illustrations of alternatives for effective integration. Both scenarios—and probably many others in between—are capable of creating an environment without barriers to the effective use of information technology for business purposes. The first scenario describes an I/S function dominated by end-user control; the second describes a well-managed centralized approach.

The End-User Scenario. In this scenario, all computing would be run by users. The central I/S department, if there is a department at all, would be reduced to a small staff responsible largely for facilities management. The four activities would be performed as follows.

I/S *planning* would be a responsibility of end-user departments and would become completely integrated into business planning. There would be no central or aggregate I/S planning per se at either a divisional or a corporate level. Both applications and architecture planning would be carried out by individuals, work groups, and departments in light of their specific business objectives. Each department would have its own systems architecture (with communications to other groups, departments, and locations as necessary). Users might occasionally ask for planning help from a small staff of in-house specialists, though they might be just as likely to use outside consultants.

The same individuals, work groups, or departments would do their own *applications development*. Most development would involve prototyping and fourth- and fifth-generation languages. Documentation would be for user purposes and tailored to each application. Many users would know how to program. Each department would have a list of preferred development tools and would employ at least one I/S technical expert. If an application that cut across departments or locations were needed, a task force representing the various locations would be formed to carry out the system development.

Operations would be the responsibility of the user/developer. User departments would hire an I/S professional staff to operate their computing facilities. Some might contract with the small I/S department to operate their mainframes. For large and complex systems, users would often rely on outside providers (such as service bureaus).

Technical services would also be located primarily within the user departments or work groups. Most departments would have formal or informal support personnel. They would issue their own guidelines for systems development and operations, occasionally asking for an I/S professional's help, but working toward self-sufficiency. An I/S professional on the corporate staff might advise the dispersed I/S function, providing a list of outside I/S professionals available for consultation when necessary. (Many of these outside professionals, by the way, would probably be former in-house I/S specialists.)

We know of no large organization today that fits this extreme end-user–dominated scenario. The scenario is plausible, however, if one extends current trends of end-user control in many of today's businesses. It assumes that the functions of many large systems can be performed by groups of small systems and that large databases can be broken into small, local databases. It also assumes that users acting in their own individual interests are simultaneously acting in the interest of the whole organization. Many observers see this outcome as inevitable.

There is no doubt that such a scenario creates an I/S function integrated with the business.

The Managed-Utility Scenario. By contrast, the second scenario arrives at integration quite differently. It is based on the premise that some information technology applications are so strategic and essential to the competitive positions of their organizations that they should be managed from the top down. Proponents of this approach will argue that strategic applications are too important to be left to "amateurs."

Planning responsibilities in this scenario would stay at the corporate and divisional levels. A corporate I/S manager would be responsible for approving major investments in technological infrastructure. Such a manager would be an active member of the top management team and an active participant in the translation of business plans and initiatives into information technology implications. User management would continue to identify applications, but I/S, as consultants to the business units, would also point out the need and value of applications. I/S professionals would compare new application proposals with existing application inventories to avoid duplications. Most important, I/S would be responsible for thorough, long-range architecture planning.

Most *applications development* work would be done by a central unit of I/S professionals. Users would probably continue to be responsible for small, individual applications; however, users would be encouraged by I/S guidelines and incentives to use only I/S-approved technology. Applications requiring more than a specified expenditure would require approval from I/S. A centralized development staff would report to the corporate manager of systems development; they would consult a user task force for needed input.

The central I/S group would also be responsible for *operations*. The corporate mainframe and a distributed network would both be operated by corporate or divisional I/S. Of course users would still operate personal computers, but they would be encouraged to follow guidelines supplied by I/S. In this scenario, I/S might even conduct occasional operational audits to ensure good operational practices on personal computers.

Technical services would be provided by a skilled team of internal consultants working as analysts within a central I/S department. An Information Center would provide users with assistance regarding development or operations of personal applications.

The managed-utility scenario derives its appeal from the sense of control and the view that a central utility can integrate the I/S function into the overall business strategy. If the end-user scenario has the strength of putting the creative power of computing in the hands of all users, the managed utility promises control and responsiveness for I/S as an instrument of corporate and divisional strategy. It runs the risk of alienating end-users, however. But for some organizations the advantages of central control will outweigh the disadvantage of loss of user involvement.

Though extremes, the scenarios present possible visions for the future.

> The key to achieving integration should be clear: the transition from the period of Domain II to Domain III will require top management arbitration and some major judgments regarding the ownership of the I/S function.

It has been a thesis of this book that such judgments are imminent in most large organizations and that, as a result, executives must develop a greater understanding of and comfort with the I/S function. The next chapter suggests specific actions for top executives and for I/S managers.

Notes

1. This definition and the discussion of integration are based in part on research by the Index Group, Inc., and Hammer and Company in PRISM (Partnership for Research in Information Systems Management), *Integrating Information Systems and the Business—Final Report* (Cambridge, Mass.: Index Group, Inc., and Hammer and Company, December 1986).
2. The historical treatment of domains is based in part on Rockart's "Eras" of computing. See J. F. Rockart and M. S. Scott Morton, "Implications of Changes in Information Technology for Corporate Strategy," *Interfaces* (January–February 1984):84–95.
3. The vendors divided the market in much the same way. IBM specialized in finance systems. Hewlett Packard and later Digital Equipment Corporation specialized in scientific and engineering systems, while Wang Laboratories supplied word processing to automate office procedures.
4. PRISM, *Integrating Information Systems and the Business.*

8
Taking Action: Transforming Businesses and Careers

W e have described the Information Imperative and its implications for businesses, executives, and the information systems function. We have encouraged readers—both senior executives and I/S professionals—to use the Benefit/Beneficiary Matrix as a framework for understanding and assessing the areas of applications of information technology. Now it is time to summarize what actions managers can take to avert the threats and take advantage of the opportunities presented by the Information Imperative.

In the first part of chapter 8 we will speak directly to business executives, both top executives and functional managers, recommending ways to effect positive changes in their industries, their organizations, their functional units, and their own careers. In the second part of the chapter, we will address I/S professionals, particularly I/S executives, suggesting what they can do to accept and lead the inevitable changes in their function and in their careers.

While this chapter suggests numerous action steps, we must caution readers that there is no one specific prescription or recipe appropriate for all businesses, functional units, or individuals. The suggestions are intended to provide general guidance and direction but not rules; they can help readers to sift through and sort out the many variables presented by the Information Imperative and the Benefit/Beneficiary Matrix and to determine plans of action for their organizations and themselves.

Actions for Business Executives

Senior executives and functional managers should *understand* the Information Imperative and its impact on their industries, their businesses, their functional units, and their careers.

This message has appeared throughout the book. Understanding is without question the first step in taking action. In summary, there are five important ways to gain further understanding of the Information Imperative:

1. Identify the internal and external forces affecting the specific industry, business, functional units, and employees' careers, and particularly how these forces are manifested in terms of information. This process is an identification of the threats and opportunities of the Information Imperative in a specific environment.

2. Understand that information technology is both a cause of and a solution to problems presented by the Information Imperative. Learn how others are using information technology. Become more familiar with the technology; for example, use a personal computer at home or at work.

3. Understand and treat information and information technology as a resource.

4. Understand that executives and managers must manage information in an aggressive way, using it to guide and often to redefine the business or business unit to compete more effectively in today's marketplace.

5. Understand how information technology can be used (and is already being used) to transform industries, businesses, business units, and personal careers. Look for patterns and parallels *outside* the environment of immediate relevance.

The Benefit/Beneficiary Matrix aids understanding, suggesting where information and information technology can help confront the threats and take advantage of the opportunities posed by the Information Imperative.

> Senior executives and functional managers should lead in setting a technology strategy and should foster conditions conducive to finding creative solutions.

Obviously, understanding is not enough. Senior executives and business unit managers must become the leaders of implementation and change. Most important, these managers should insist that information technology planning be linked to the firm's strategic planning (and vice versa), and they should be active in the planning process.

The Benefit/Beneficiary Matrix offers a framework for sorting out and assigning priorities to the business benefits of information technology (increasing efficiency, improving effectiveness, achieving transformation) and for targeting specific beneficiaries (individuals, functional units, the organization as a whole). In addition, it is a valuable tool for examining an organization's current portfolio of information technology applications and for deciding what the future portfolio should look like.

Successful leadership requires finding specific promising business opportunities, including the new types of opportunities made possible by the information technology revolution. Education and understanding help suggest where to look. In chapters 3 and 4 we suggested three categories to consider when looking for solutions:

1. Redefining the marketplace offering or the functional department's output

2. Restructuring connections—between internal and external constituencies or internally among different departments

3. Redesigning business and management processes for the business as a whole or for functional units

In each of the three solution categories, executives should:

Target specific actions to specific threats or opportunities; across-the-board tactics often do not work

Be flexible—be able to adapt and then adapt again

Integrate activities—piecemeal solutions executed department by department will not suffice for today's most difficult challenges

Select priorities carefully, identifying the most promising actions before implementing solutions

Explore and manage the roles of information and information technology in implementation

Senior executives and business unit managers should be prepared to act in innovative ways to shape and encourage new approaches to business and new mind-sets and culture. They should become informed visionaries, champions of change, and architects of the change process.

We have argued that many ideas for creative solutions will bubble up within the organization. For many organizations, inspiring innovation from the bottom up will be difficult. The degree of change will depend on the organization, its industry, the I/S mission, how the I/S functional activities get distributed into I/S organizational units, and what kind of relationships exist between managers and their employees. All companies will not—and should not—look the same. But many companies will find it necessary to make fundamental changes in their corporate culture in order to encourage innovative use of information and technology (and innovation in general). Here are five suggestions:

1. Foster a culture of innovation in which virtually every employee is almost an R&D center supported by I/S.

2. Create a climate of trust that information technology will be used for individual and organizational benefit, not harm.

3. Accept and/or develop a supportive attitude about taking risk.

4. Be open to experimentation: be willing to try creative ideas for using information technology regardless of where in the organization they arise.

5. Commit to having technology available throughout the organization.

Senior executives should establish a clear vision for the future and communicate that vision throughout the entire organization. It is also important that they become role models for the vision, "living the vision" by their own examples. Often, they should "get their hands dirty" experimenting with technology. We believe that every executive should work (directly or through assistants) to experience personal computing, executive information systems, and other systems that transform the performance of individuals, functional units, and/or the business a whole.

Actions for I/S Professionals

It is generally recognized that I/S managers today cannot continue as if permanently rooted in the Domain II period. Just about everyone has an opinion about what they should do:

Users want less interference, more control.

Senior managers want an executive-level I/S decisionmaker who knows the business and understands technology.

I/S specialists want leadership and direction plus reassurances that their jobs are secure.

In the middle, swamped by all these opinions, stands a sometimes paralyzed I/S executive, stuck between doing things "the way we have always done them" and making what seem like risky and uncomfortable changes.

I/S executives must recognize and accept the fact that I/S as it is today will change, or it will not survive. The presence of a strong central I/S unit, albeit different from what we now know, is necessary in order for most organizations to achieve integration. In the remainder of this chapter, we will suggest strategies for I/S for coping with change and providing leadership. We will also return to the four key I/S functional activities (planning, applications development, operations, and technical service and support) to highlight the implications of anticipated changes and to suggest the "shoulds" for I/S executives and their organizations in responding to the Information Imperative.

As an immediate priority, I/S professionals should work to overcome their resistance to change.

No one questions that the I/S function is in transition. But, as was discussed in chapter 7, how it will change and which scenario will work for a particular organization are unclear at this time. It is clear, however, that I/S professionals can no longer resist or ignore the need to change.

One key to success is the I/S executive's ability to gain the confidence of senior and line management. The confidence gap must be closed, by one or all of the following means:

Mutual education and familiarity

Responsiveness to user needs

A shift of job or company

Another key to success will be an I/S executive's ability to manage multiple changes simultaneously, recognizing their inter-relationships. Transformation of the I/S function and of I/S executives' careers will require management of multiple changes in the business, the technology, and the I/S organizational structure. As we see it, the I/S executive of the future will be instrumental in guiding technology-enabled transformation (for the business, a functional unit or an individual) to ensure success. This role may imply a Chief Information Officer or "CIO" position, with or without line responsibility for an I/S unit. One of the first action steps for I/S executives is for them to examine their own attitudes about change, seeking a better understanding and acceptance of the impact of change on their own careers and on the job responsibilities and career opportunities of their subordinates.

Organizations committed to moving into the Domain III era can help I/S professionals address the issue of change by clarifying career paths and offering business training and career counseling. In addition, organizations can set up programs (such as internships and multidisciplinary teams) to bring users and I/S professionals together more often, encouraging each group to learn from the other.

What are the implications for I/S as each of the four key I/S activities moves in the Domain III era?

> I/S should change planning in light of the push for Domain III applications and the continuing growth of end-user computing.

Today's I/S organization often has comprehensive planning disciplines covering many basic I/S activities (systems development, capacity planning, facilities planning, and so forth) but falls short in planning for end-user computing. New disciplines, including user participation and more appropriate justification measures, are required.[1]

Often even more important than planning for end-user computing is the need for I/S to be more actively involved in overall business planning. The I/S executive should become an integral part of the top management decision-making team. Most experts agree that traditional (often bottom-up) planning approaches are overly reactive and place most organizations at a competitive disadvantage with regard to information technology. People historically responsible for generating systems requests rarely linked those requests to overall strategy; in fact, too often the people who generated those requests did not understand the overall business direction adequately.

I/S executives should assume much of the responsibility for change. They should work actively to understand the overall business strategy and the varying emphases of different segments of the business. If, for example, Division A focuses on low-cost manufacturing, Division B emphasizes the introduction of new products, Divisions C and D are targets for consolidation, and Division E is to be sold, then each division would have different applications needs. Even in a company with only one line of business, the I/S executive should recognize the role and priorities of each of the different functions (product development, manufacturing, sales, and so on) before determining the appropriate levels of information technology support and investment. While I/S executives are responsible for delivering technical solutions, they also share in the responsibility for defining business goals and ensuring that the appropriate technology is employed to achieve those goals.

Finally, we believe that I/S executives should also spearhead the move to find better measures of benefits and risks of information technology applications. Traditional risk measures include size (larger projects are riskier than smaller ones), the organization's familiarity with the proposed technology (a system is riskier when a company is not accustomed to the technology), how well the problem has been defined (a poorly defined problem carries more risk), and the effectiveness of the multidisciplinary team (a poorly organized or ineffectual team is riskier). The traditional benefit measure is headcount and keystroke.

Transformation in the age of Domain III applications requires broader measures for both benefits and risks. System benefits will relate to broader business values such as increasing market share, creating barriers to competitive entry, or establishing competitive advantage through improved relationships with suppliers or customers. Often accompanying the promise of larger benefits will be greater risks. The risk of poor timing (allowing the competition to move first) will increase in importance. So will the risk of poor implementation, in part because of the increased business importance of Domain III applications and because those applications will often reach outside the business to impact customers and others.

> I/S should see that applications development takes full advantage of new tools and techniques.

In the future, at least three major types of applications will be required: (1) large transaction-processing applications to augment and in some cases replace today's systems (2) executive information systems, and (3) other end-user applications. These types of applications differ dramatically; they require different skills, tools, techniques, and project management controls. As was discussed in chapter 7, development of Domain III applications will require a combination of the traditional structured approach with prototyping and advanced tools such as fourth- and fifth-generation programming languages and graphics software. In particular, systems development methodologies that provide a rapid, interative, systems-building environment will be required to meet the timing and learning needs of end users. Table 8–1 provides an outline of differences in approaches to the management of transaction processing and executive information systems.

Most of today's I/S departments are structured according to

Table 8–1

Typical Characteristics of Transaction Processing and Executive Information Systems

Characteristic	Transaction Processing	Executive Information
Team size	Large	Small
Expected life	Long	Short
Primary language	Traditional (third generation)	Fourth or fifth generation
Development approach	Structured	Evolutionary
User participation	Large amount, lower level	Medium amount, higher level
Programming effort required	Large	Small
Timeframe covered by the application	Current and recent past	Past and future

the business functions they support (one unit to support finance, another unit to support manufacturing, and so forth). "Focused factories," each providing one delivery mechanism (such as transaction systems or executive information systems), may offer an attractive alternative to today's structure.

Transformation also requires attention to effective productivity measurement within I/S. I/S organizations have been reluctant to introduce productivity measures in the past. We believe that such measures will be useful in providing baseline productivity information and especially in determining the effectiveness of specific tools and techniques. A growing number of commercial products aimed at improving productivity have already been introduced, and many I/S departments are also developing additional tools internally. We believe some of the major areas of emphasis should be:

Automated design tools

Code generators

Reusable codes

Automated documentation systems

Special-use programming languages

We also recommend that I/S executives monitor new technology options, such as voice recognition, touch screens, expert systems, and many others, and that they consider the applicability of such advances to systems development activities.

> I/S should adapt project management to accommodate Domain III applications in general and, in particular, executive information systems and end-user applications.

In particular, we believe that I/S executives should elicit the support of an active and committed champion for each Domain II or III application. In general, the champion should be a senior manager. When an application requires integration across functional units (as many Domain III systems do), the champion should often have authority that spans the affected functions. In addition, we recommend using multidisciplinary teams made up of representatives from a range of functional units to establish the intended result of the application and to help guide its definition.

> I/S should emphasize technical service and support activities for Domain III applications.

Throughout this book we have suggested that the I/S executives, along with other employees in the I/S organization, should become more visible and accessible to users and senior executives alike. I/S's role is, at least for some applications, shifting from that of "doer" to that of facilitator. In their new role, I/S executives and their teams can and should educate senior executives and key users to the power and advanced capabilities of technology. They should recognize "opinion leaders" within the organization and support the technological development of those leaders.

One very practical change that might help ease tensions between I/S and users would be to reinstitute the role of business systems analyst. At one time, individuals in such jobs served as the interface between users and the systems organization. For a variety of reasons, these positions have generally been eliminated or refocused to make them more technical. Often the interface responsibilities have effectively shifted to users. We believe we have come full circle and that there is once again a need for the business systems analyst.

Finally, we suggest that the I/S executive take another look at the technical support unit, often working to do more with less. Technical support is vital for cost-effective end-user computing. Technical scanning is increasing in importance. But support costs can grow swiftly and require careful management.

> I/S should work to explore and master advances in technology.

The proliferation of new technology and new options presented by technology require that I/S provide more education both inside and outside the I/S department. For example, one need is for improved knowledge and skills relating to the integration of computers and telecommunications. The integration of voice with data communications and of computers with communications creates new challenges for the I/S department. I/S professionals have extensive knowledge and skills in computers and sometimes in data communications, but not in voice. I/S executives will need to add new skills in the voice communications field and master the issues of integration.

Another of I/S's major challenges is bringing the organization's data resource under control. We have argued that information is a key corporate resource. Many leaders in I/S believe that we do not today have satisfactory answers to the problem of data architecture (of how to structure and manage the basic data resource). Finding answers is a critical challenge for transformation in the era of Domain III applications.

The Imperative for I/S

> In summary, we propose, that I/S executives understand and provide leadership, not necessarily ownership, of all I/S functions within an organization.

We believe this action is appropriate regardless of whether the I/S function is centralized or decentralized and regardless of whether or not the I/S unit has total responsibility for the I/S function. We urge I/S executives to embrace leadership roles enthusiastically. In doing so, they will help their organizations use the right technologies and applications to maximize business benefit. In the process, they will further their own careers and professional development.

Note

1. A new approach to I/S planning is described in PRISM, *Information Systems Planning in the Contemporary Environment* (Cambridge, Mass.: Index Group, Inc., and Hammer and Company, December 1986).

Index

About the Authors

Cyrus F. (Chuck) Gibson is Senior Vice President of the Index Group. His client service areas include custom executive education on information technology and consulting on the organizational and behavioral impacts of technology. Before joining the Index Group in 1978, he was an associate professor at the Harvard Business School, where for nine years he taught organizational behavior in the M.B.A. program and general management and information systems management in executive education programs. He is author of the book *Managing Organizational Behavior,* coauthor of "Managing the Four Stages of EDP Growth," and author or coauthor of numerous other articles and cases on the implementation of systems and organizational change. Dr. Gibson holds a bachelor's degree in mechanical engineering from Yale University, an M.B.A. from Harvard University, and a Ph.D. in organizational studies from the Sloan School of M.I.T.

Barbara Bund Jackson is an independent consultant and teacher in marketing based in Belmont, Massachusetts. She works with clients on market research and analysis, the formulation of marketing strategies, and the implementation of those strategies and tactics. Previously, as a vice president of the Index Group, she headed the firm's marketing practice, working with some clients to use information and technology in marketing and sales and with others to market information technology products and services. Before joining the Index Group, she spent eleven years on the faculty of the Harvard Business School, teaching marketing, industrial marketing, and the use of computers and related tools in business. Dr. Jackson's most recent publications are *Winning and Keeping Industrial Customers* (Lexington Books) and "Build Customer Relationships That Last" *(Harvard Business Review)*. She holds an A.B. and a Ph.D. from Harvard.

About the Index Group

Index Group, Inc., is a leading international management consulting firm that helps organizations use information and information technology for business success. Founded in 1969, the firm now has offices in Boston, Los Angeles, and London. The firm is known for its pioneering work in the areas of strategic systems, decision and management support systems, the critical success factors (CSFs) process, systems prototyping, management education, and the management of change. Offering a blend of consulting, education, and research services, the Index Group enables its clients to tackle the challenges and problems presented by information and information technology.